Dead Reckoning

Confronting the Crisis
in Pacific Fisheries

Terry Glavin

The David Suzuki Foundation

GREYSTONE BOOKS
Douglas & McIntyre
Vancouver / Toronto

This book is printed on 100% tree-free, non-chlorinated Kenaf paper.

Greystone Books
A division of Douglas & McIntyre Ltd.
1615 Venables Street
Vancouver, British Columbia
V5L 2H1

The David Suzuki Foundation
219 – 2211 West 4th Avenue
Vancouver, British Columbia
V6K 4S2

Canadian Cataloguing in Publication Data

Glavin, Terry, 1955-
   Dead reckoning

   ISBN 1-55054-487-X

1. Fisheries—Pacific Coast (North America) 2. Fishery conservation—Pacific Coast (North America) I. Title.
SH214.4.G62   639.2'09164'3   C95-911230-8

Editing by Barbara Pulling
Cover photograph by Flip Nicklin/First Light
Cover design by DesignGeist
Typeset by Brenda and Neil West, BN Typographics West
Printed and bound in Canada by Friesens
Printed on 100% tree-free, non-chlorinated Kenaf paper

The publisher gratefully acknowledges the assistance of the Canada Council and of the British Columbia Ministry of Tourism, Small Business and Culture.

Contents

Foreword

THE DAVID SUZUKI FOUNDATION WAS ESTABLISHED IN 1990 TO FIND concrete ways of achieving a sustainable balance between humankind's social, economic and ecological needs. We communicate our findings to all levels of society and develop strategies for change with local grassroots individuals and groups, for we believe that change does not trickle down from the top but rises from local communities to government and corporate levels. We also believe that there is still time to change the way we live and to create a brighter future.

We are impelled by the sense of urgency in the Warning to Humanity signed by 1600 senior scientists from seventy-one countries and over half of all living Nobel Prize winners: "Human beings and the natural world are on a collision course ... many of our current practices put at serious risk the future that we wish ... and may so alter the world that it will be unable to sustain life in the manner that we know ... No more than one or a few decades remain before the chance to avert the threats we now confront will be lost." (The Union of Concerned Scientists, November 1992)

For decades, alarming signs of environmental degradation across the planet have been reported. Now overpopulation, toxic pollution, atmospheric change, deforestation, soil degradation and so on are directly affecting people's lives. In addition, it has become clear that the economy and social issues are inextricably intertwined with the environment.

In a mere century, humanity has been transformed into a super-species able to change the very biological and physical make-up of the Earth. Our scientific and technological prowess has allowed us to increase our consumption of energy and resources, produce pollutants and destroy ecosystems. At the same time, our shift from the countryside to the city has broken our contact with nature and freed the global economy to push for endless growth, ransacking our children's future.

As biological beings, we depend on the quality of the air, water and soil and on the diversity of life on Earth for our survival. The David Suzuki Foundation aims to document the perilous state of our ecological life-support systems while identifying those nations, regions, groups or individuals who are working to achieve sustainable living. The good news is that most experts believe that a sustainable future is possible and that there is still time to bring it about.

The David Suzuki Foundation Series is part of the communication of our work. We hope that the books in this series will illuminate both the challenges that face us and the possibilities for a sustainable future for us all.

David Suzuki
Chair, The David Suzuki Foundation

Preface

IT IS MUCH LATER THAN WE THINK.

That is the unfortunate finding any honest assessment of the state of the world's fish populations must report. Any honest assessment of Canada's Pacific fisheries will lead to the same conclusion. The deeper the inquiry, the more certain that conclusion becomes.

But it is not too late. That is the important thing. There is certainly evidence enough to be optimistic about the prospects for the restoration of marine ecosystems and fish populations, along with the restoration of sustainable, viable and healthy economies built upon them.

At its heart, this is the same contradiction that confounds conservationists the world over. The human species is only now confronting the reality of the collapse of entire ecosystems. Just as the human population grows and strains against the delicate ozone membrane that encloses the fragile planet—at this same frightening moment—there is much persuasive evidence to justify hopefulness. The discussion can get quite confusing, and "environmentalists" sometimes find themselves divided into opposing camps of optimists and pessimists. It has left the public justifiably frustrated for want of comprehensible explanations and realistic solutions.

The facts show that there are workable solutions for pragmatic recovery and a way back to equilibrium, sustainability and hope. Beyond that, what is also certain is that all of us are involved in this, whether we like it or not. This is not a ship with lifeboats.

Throughout the world, solutions are beginning to emerge, albeit painfully slowly, where it counts: in the realm of effective public policy. Some solutions to the dilemma of sustainability lie in old knowledge newly revived. Some solutions arise from entirely new intelligence about the natural systems of the planet.

There are many hands involved in this work now. I defer to these people, since they are economists, biologists, sociologists, maritime anthropologists and otherwise intelligent people, and they have some familiarity with these waters. In this book, I have tried to record their bearings as accurately as I can. Their work is discussed at length throughout, and the sources for these discussions are cited in the source notes at the end of the book, in order of their appearance in the text. These people often disagree with each other, both within their own fields of study and across disciplines. When this happens, I make some rough observations of my own from time to time, which is what writers should do in such situations.

Trawlers, seiners, trollers, gillnetters, tribal fishers and anglers routinely disagree amongst themselves about where we are and what got us here. When this happens, I have tried to pay closer attention to those things they agree upon, which can be summed up as the following: this sure is dirty weather we are passing through. We all have to take responsibility, and it's no longer good enough to leave decisions to government experts or industry experts. We'd better think hard about what we've left astern if we want to make sense of where we're at, and about what's likely looming off the bow.

That, more or less, is what dead reckoning is.

This book began with David Suzuki, Tara Cullis, Jim Fulton and all the others at the David Suzuki Foundation. Stan Persky, who knows a lot about metaphysics and things like that, is largely responsible for my involvement in the project.

Evidence for the possibility of truly "sustainable" fisheries was provided by the brilliant work of academics such as Evelyn Pinkerton and Martin Weinstein, whose findings have informed whole sections of this book. Catherine Stewart of Greenpeace Canada made available a tremendous amount of research that contributed greatly to the third chapter. The entire manuscript is significantly informed by the works of individuals such as Lester Brown, Thomas Northcote, William Ricker, Brian Riddell, Richard Beamish, Elliott Norse, Dianne Newell, Parzival Copes, Geoff Meggs, James McGoodwin, Carl Walters, Gordon Mohs, Reuben Ware, David Ellis, Michael Kew and Brian Chisholm.

Just as important was the assistance of Terry Slack, Dan Edwards, Ken Glover, Tony Pletcher and Bert Richardson. There are hundreds of people like them out there now, in the ditches and the creeks, from the Skeena to the Stillaguamish. Their wisdom, their common sense and their gumboot labour are all the evidence I need for great optimism. But this project would have failed completely were it not for the contributions of numerous dedicated employees of the Department of Fisheries and Oceans. DFO biologists and technicians assisted immeasurably with the manuscript, both in its content and in its review. It is comment enough that to name them could irreparably damage their careers.

Those who reviewed the manuscript and helped me with a variety of difficulties include David Suzuki, Jim Fulton, Stan Persky and Carl Walters. Many thanks are also due to Ben Parfitt, who is a colleague and a great friend.

The lion's share of credit, however, must go to the men and women of the fishery. They feed the rest of us. It was the patience and instruction of dozens of these people, from upriver native communities to the trawlers on the high seas, that really produced this book.

This book is for them.

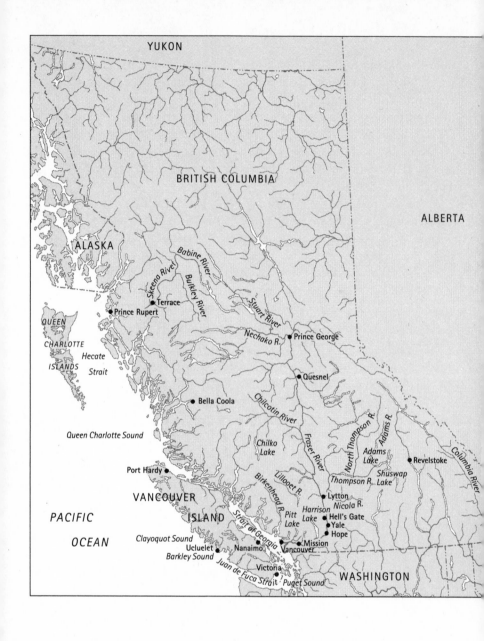

Chapter I

⚓

Who Killed Byrne Creek?

When we include ourselves as parts or belongings of the world
we are trying to preserve, then obviously we can no longer think of the
world as "the environment"—something out there around us. We can see
that our relation to the world surpasses mere connection and verges on
identity. And we can see that our right to live in this world whose
parts we are is a right that is strictly conditioned ... even if
we merely want to live, we cannot exempt use from care.

—Wendell Berry, "The Obligation of Care"

WHEN I WAS A BOY, BYRNE CREEK ROSE IN LITTLE RIVULETS THAT trickled out of a wooded ravine in Burnaby's south slope, tumbled through culverts onto "the flats" and flowed quietly through the peat bog where we poled our rafts. We hunted frogs and got spooked by the pheasants that burst from the warren of tunnels we had made through the blackberry bushes. There were cutthroat trout in the little streams; coho salmon still spawned in the creek, and old Chinese farmers put their crayfish traps underneath the overgrown bridges that crossed it here and there. There were strange little fish that swam underneath the log booms and rumours of river monsters in the Fraser's unfathomable depths. There were gillnet boats tied to rickety floats at the creek mouth, and in the summertime I used to sit under the cottonwood trees and watch the boats set their nets for sockeye up and down the Fraser River, from the old railway bridge right around the Big Bend and down towards the sea.

I didn't know then that these things would lead me out into the middle of the Pacific Ocean, to spend weeks getting up at morning's first light to pull squid and pomfret and albatross from a driftnet set alongside the nets of Asian driftnet ships, doing my tiny bit in a Canada-U.S. research effort into the effects of high-seas driftnet fishing. I didn't know then that these things would lead me to stand and watch as Stl'atl'imx fishermen, ropes tied to their waists, waded with their dipnets into the Fraser's violent cataracts to precisely the right crevice, to haul bright silver chinook salmon as big as children from the depths of the thundering river while federal fisheries officers watched through their binoculars from the hills far above. I didn't know then that I would spend much of my adult life writing about fish, and learning about the ways people catch fish and what fish mean to the planet.

As a boy, I didn't know that the reach of the river where we played on log booms had been fished for centuries by Musqueam and Kwantlen and Cowichan fishing families. And as I played along the tree-lined bank of Byrne Creek in the early 1960s, I didn't know that those were the last days coho salmon would ever spawn there.

Out in the river, fishing from one of the gillnetters, was Terry Slack, whose father and grandfather had fished the North Arm of the Fraser before him. On a summer day in 1962, long before subdivisions enclosed the ravines, and industrial parks and a golf course and a highway covered the peat bog, long before the new flood-control pump went in at the mouth of Byrne Creek, Terry Slack pulled from his net a big fish, known in the scientific literature as *Acipenser medirostris*, known along the river simply as a green sturgeon.

Green sturgeon are strange animals. Seven-footers were not uncommon at the mouth of the river around Sturgeon Bank (so named by Captain George Vancouver, who acquired some sturgeon in trade there from the Musqueam people in 1792). One of three species of saltwater sturgeon on the planet and the only sea sturgeon in the Pacific, the green sturgeon was the last of the world's acipenseriformes to reveal itself to Western science. It wasn't even officially recorded on the British Columbia coast until 1908. Back in the 1930s, Terry's grandfather, John Slack, used to pull as many as half a dozen green sturgeon from each set of his gillnet at the mouth of the North Arm. But every year they were growing fewer, and that summer day in 1962 was the last time Terry Slack would ever see a green sturgeon in his nets.

I didn't meet Terry Slack until more than thirty years had passed.

By then, he had spent years wondering why he and his fellow gillnetters weren't seeing green sturgeon any more. He had been calling every government fisheries office and scientific agency he could think of. Nobody knew a thing about green sturgeon. So after a while he posted notices down on the fish docks at Steveston and False Creek, asking fellow fishermen to let him know when they had last caught one, or even heard about one. He talked it up at fishermen's meetings, and *Westcoast Fisherman* magazine printed a notice asking anyone who had caught a green sturgeon to call Slack and tell him about it.

Slack was swamped with calls. When it was all over, as far as he could tell it had been four years since a green sturgeon had been seen in the Fraser River. In 1991, one green sturgeon had turned up in a fisheries department test-fishery net at Maple Ridge, and that same year there was a report of a green sturgeon coming up in a gillnet during a pink salmon opening at the mouth of the river, about four miles off Steveston.

When I first met Terry Slack, the West Coast's fisheries were in a "missing fish" uproar. There were official investigations, public hearings, allegations and counter-allegations, protests and counter-protests, and green sturgeon weren't all that important any more. Byrne Creek wasn't important, either. It was just one of dozens and dozens of creeks in Burnaby and Vancouver and Richmond and Surrey and Langley that had already gone, making way for other things. All the shouting and the controversy were made more urgent by what had happened on the East Coast. All the talk was about the collapse of the Newfoundland fisheries, and whether anything like it could happen on the West Coast. Most of the experts said the same thing could not happen here.

Which is true enough.

Unlike the North Atlantic cod fishery, the Pacific fisheries depend upon hundreds of distinct fish populations that fall within several broad taxonomic classifications and comprise creatures as diverse as geoducks, herring, sea cucumbers, salmon, halibut, crab and deep-dwelling rockfish. Even the sockeye salmon fished by Fraser River gillnetters like Slack are made up of dozens of distinct stocks, each with its own unique way of behaving, each run bound for its own suite of spawning streams around lakes as close to saltwater as Pitt Lake, only a few kilometres upriver, and as far away as Stuart Lake, hundreds of kilometres to the north. Unlike the Grand Banks and Georges Bank of the Canadian and U.S. East Coast, the primary resources of the

West Coast's fisheries are not confined to a single, common range. So our fisheries are less likely to be wiped out by one horrific, death-dealing blast.

Instead, our fisheries are more likely to die a long and painful death, fishery by fishery, stock by stock, species by species, creek by creek. A complete fisheries collapse on this coast would more likely be the consequence of increments, each collapse more or less confined to a single population within a single species, each closure affecting small groups of people whose ways of life are now so different from the mainstream that their protests would melt away into what are already incomprehensible disputes and controversies within the Pacific fishery.

For most British Columbians, it must be a baffling affair, this melodrama that unfolds almost every night on the television news and every morning in the newspapers. There are fish-dependent communities that appear to be at each other's throats, while other fishing groups behave as though they are besieged by the world itself. Fishers stage anti-government protests as often as they go fishing. They blame one another, they blame federal fisheries managers, they blame Americans. Sports fishers say it's the commercial fleet's fault. Much of the commercial fleet says the Indians are wrecking everything. There are native leaders who blame the white people. Sometimes the villain is something called "El Niño," which is reported to be a phenomenon of some sort in the ocean. Sometimes it's the seals or the sea lions that are taking all our fish.

The kind of news coverage that characterized the 1995 fishing season in British Columbia was typical. There was generous attention paid to events some fishing industry activists staged for the media in their ongoing disputes with native fishers on the Fraser River, and there was a flurry of broadcasts and news controversies arising from a media campaign waged by the sportsfishing lobby to win a greater share of the catch at the commercial industry's expense. But the story that dominated the season was one with great public relations prospects for Canadian politicians: the story about a beleaguered Canadian government teaming up with American Indians and conservationists to restrain reckless overfishing by Alaskan trollers.

After having successfully cast Spanish trawlers in the bad-guy role in a similar public relations effort on Canada's East Coast a few months earlier, federal Fisheries Minister Brian Tobin orchestrated a series of "news" events over the summer of 1995 in his campaign to blame

Alaska for threats to B.C.'s endangered chinook stocks. The dispute centred on Alaska's insistence on a chinook allocation to its troll fleet of 230,000 fish, most of which spawn in B.C., Washington and Oregon rivers. In the absence of an agreement under the Pacific Salmon Treaty, Canadian scientists—with the support of B.C.'s commercial, recreational and aboriginal fishing communities—called for a scaled-back Alaskan fishery to keep the chinook catch below 138,000 fish. There were court skirmishes and press conferences and "goodwill" missions to Alaska, and Alaskan authorities remained isolated throughout the season. Even the U.S. National Marine Fisheries Service ended up siding with Canada.

Long after the press conferences and media events were over, whatever chinook were "saved" by interventions against Alaska ended up getting caught in Canadian waters anyway. B.C.'s northern commercial trollers were handed new chinook openings after Alaska's troll fishery was scaled back, and B.C.'s sports fleet hit chinook stocks even harder. Dan Edwards, the forty-five-year-old skipper of the commercial troller *Tidewinder* and resident of the fishing village of Ucluelet on Vancouver Island's west coast, watched it all happen in a quiet rage. He wasn't alone. He was joined in his anger by the West Coast Sustainability Association, the Pacific Seafood Council, the Pacific Rim Fish and Game Association and the Nuu-chah-nulth Tribal Council. They could do nothing more than sit and watch as an armada of sportsfishing boats were allowed to continue catching endangered chinook from dozens of Vancouver Island rivers. While a handful of the sportsfishing boats Edwards watched were local, a lot of them were American sportsfishing yachts, and others were from the fleets of catcher boats dispatched from transient charter operations. Despite Canadian protests against Alaska, the Canadian government authorized continued catches of chinook salmon from more than sixty endangered runs around Esperanza Inlet, Barkley Sound, Kyuquot Sound, Clayoquot Sound and Nootka Sound. In the decade preceding the Alaska rumpus, several spawning streams on Vancouver Island's west coast—many of them just like Byrne Creek in its final days as a salmon stream—had lost all but a few dozen spawners.

But Edwards's protest seemed to disappear under more dramatic headlines. It was hard to make sense of because it didn't fit a plot played out by the more recognizable characters we have established for ourselves: native poachers or reckless Alaskans or incompetent federal bureaucrats. So Alaska remained the villain in the news for most of the fishing season. Greenpeace Canada, the Sierra Club and other B.C. conservationist organizations blasted Canada's domestic fishing plans as a "recipe for further stock extinctions," but their warnings went largely unnoticed.

When the World Wildlife Fund urged a catch-and-release rule in the 1995 sports fishery for Strait of Georgia chinook salmon, theirs was also a story too complicated to warrant more than one or two spots on the evening news. As recently as the 1970s, chinook salmon spawning in the streams and rivers around the Strait of Georgia were counted in the hundreds of thousands. Yet according to the Canadian Department of Fisheries and Oceans' own stock assessments, the number of Strait of Georgia chinook spawners between 1985 and 1995 had not once exceeded 15,000 fish, and in 1987, only 4,500 chinook spawners were counted in all of the strait's streams. Nonetheless, in 1995, B.C.'s influential sportsfishing lobby kept the fishery open all season.

Strait of Georgia coho salmon are under the same kind of pressure, with most populations on the decline. By 1985, commercial and sports harvests routinely produced coho harvest rates "generally exceeding 70 per cent and reaching as high as 85 per cent," according to DFO's 1995 South Coast Coho Integrated Management Plan report. (A "harvest rate" of 70 per cent means the combined catch takes seven out of every ten adult fish.) This rate of fishing had been going on for several years by then, despite the warnings of the federal government's own biologists. From 1976 onward, the catch of Strait of Georgia coho had declined by about 66,000 fish per year. Escapement (the number of fish that survived the fishing gauntlet to spawn in their home streams) had fallen by half, and spawning had become concentrated in fewer and fewer streams, reducing genetic variability. Still, by 1995 conservation measures included only a ban on the commercial troll catch of coho in Strait of Georgia waters—a catch that had become largely insignificant by the early 1990s anyway—along with minimum size limits and a daily "bag limit" reduction from four to two fish imposed on sports fishers, who take most of the coho catch. Most conservationists and DFO biologists regarded the measures as wholly insufficient and urged instead a ban on all coho harvests. The fishing continued.

Following a recent migratory pattern believed to be linked to rising water temperatures and declining feed in inshore waters, the Strait of Georgia coho appeared to move out en masse to the west coast of

Vancouver Island in 1995, making them vulnerable to the "outside" troll fleet by the end of July. The cohos' only migratory paths homeward were blocked by the trollers, seine boats and gillnet boats on Vancouver Island's west coast, in Johnstone Strait and the Strait of Juan de Fuca—all preparing to set their nets in sockeye and pink fisheries that are known to kill coho salmon in high numbers. Conservationists were reduced to issuing public appeals to anglers throughout the Georgia Strait to release alive any coho they caught.

On the Fraser River, initial spawning-escapement goals for sockeye salmon were abandoned in favour of much more modest goals as the commercial fishing industry demanded more openings, casting their demands as protests against "native-only" fisheries. The commercial fleet got its openings, and the Fraser River tribes failed to come close to catching their pre-season allocations. All the while, far removed from the headlines, the sockeye were suffering yet another year of overfishing.

The villains change, but the declines continue. On Canada's West Coast today, many distinct fish populations are already as "commercially extinct" as the North Atlantic cod. Many fish populations have been extirpated from much of their former range. In the case of Fraser River pink salmon, which was once dispersed throughout the Fraser basin, 90 per cent of the basin's populations now spawn only in the final two hundred kilometres of the river below Yale.

Many local fisheries throughout the B.C. coast are already long gone and long forgotten. Up to 75 tons of surf smelts were harvested annually from the waters around the mouth of the Fraser River until that fishery collapsed in the early 1970s. Already, it has been several decades since halibut were hauled in by hand line in English Bay for sale in the downtown markets. It has been a long time since salmon trollers bobbed around Burrard Inlet, every evening from early spring to late fall, waiting for the nine o'clock gun to sound the end of the day's fishing.

In fact, British Columbia's fisheries are headed towards precisely the same place the Newfoundland fishery has ended up. But since our declines are comparatively piecemeal and inch-by-inch, hardly anyone notices. At this rate, they will soon be the concern only of academics and historians, just as, for most Vancouverites, the boom of the nine o'clock gun is just some old tradition. For most of us, the spectacular abundance of fish that supported human societies on this coast for 10,000 years will gradually be forgotten, in the same way that we have already forgotten so many things, some as small as Byrne Creek, some as vast as the herds of humpback whales that once flourished in the Strait of Georgia. The last of the Georgia Strait humpbacks, forty all told, were slaughtered between November 15 and December 30, 1907, by the Pacific Whaling Company ship *Orion*, when its crewmen fired the ship's harpoon cannon into the remaining humpbacks off Comox and the mouth of the Fraser River. The *Orion* towed the dead whales to Piper's Lagoon, just outside Nanaimo, for rendering into oil. And then it was over. Despite warnings, the whalers reassured everyone that there were lots of whales, and then suddenly the whales were gone.

In the same way, eastern fishing companies and far too many DFO scientists said there were lots of cod on the Grand Banks, and now the cod are gone. The collapse of the stocks off Newfoundland and Labrador was foreseen by inshore fishermen, whose nets had been coming up empty for years. Some fisheries scientists had also been issuing warnings long before the fishery was finally shut down. While the political clout of the cod fishery's major harvesting and processing firms should not be underestimated, the scientific models fisheries managers relied upon to assess the impact of increasingly complex and increasingly "catch-efficient" fisheries simply failed to show the sort of precision that politicians had come to demand to justify unpopular fishing closures. So little is understood about the ocean, stock productivity and fish population dynamics that a long-term, gradual decline in ocean temperatures in the North Atlantic—which appeared to produce a rising trend in cod stock recruitment (the rate at which a harvested fish population can replenish itself)—may have masked the real impact of increased cod catches. Whatever the role these long-term, little-understood cycles may have played in DFO's stock-assessment errors, the department grossly miscalculated the size and resilience of the northern cod stocks. But nobody knows for certain what happened.

The collapse of the northern cod fishery has few parallels in the history of the planet's fisheries. The fishery began centuries before the arrival of Europeans, but even the European fishery on the Grand Banks almost certainly began before Columbus, perhaps as early as 1481. The fishery was the mainstay of the Newfoundland economy for centuries. It supported tens of thousands of jobs, and even in its last years it was worth about half a billion dollars annually. Overfishing in the 1960s and early 1970s (offshore trawlers harvested an unprecedented 800,000 tons of fish in 1968) prompted Canada to extend its two-hundred-mile limit to enclose the fishing grounds in 1977, and cod populations became the subject of intense scientific scrutiny. Still, none of this scrutiny made any difference in the end. In the case of northern cod, it is probably true enough to say we didn't know the first thing about what we were doing.

Three years into the East Coast moratorium on cod fishing, there was still no evidence that the stocks were recovering. More than forty thousand workers remained on the dole, and stock assessment surveys showed that Grand Banks cod populations were at about 1 per cent of what they had been only a few short years earlier. North Atlantic populations of flounder, white hake and ocean perch are also disappearing, and their decline has forced further fishing closures on both Newfoundland coasts, in the Gulf of Saint Lawrence and off Nova Scotia's southern coast. Pollock and flatfish quotas have been slashed as well.

South of the border, along the U.S. East Coast, the Georges Bank fishing grounds have undergone a similar process of rapid marine desertification. Between 1982 and 1993, annual landings of Georges Bank cod fell from 57.2 million tons to 23.1 million tons. Georges Bank haddock landings dropped from 18.2 million tons to 4.5 million tons, and yellowtail flounder landings fell from 10.5 million tons to 2.1 million tons. Two weeks before Christmas, 1994, about six thousand square miles of ocean, formerly among the most productive fishing grounds on the planet, were closed to fishing. For the first time in three centuries, boats stayed tied to their picturesque docks in New Bedford and Gloucester and dozens of other New England towns. Their skippers were left with little hope of ever going fishing again. Directly to the south of Georges Bank, on Chesapeake Bay—once another of the world's most productive fishing grounds—the annual oyster harvest had steadily declined to 1,000 tons by 1993, a catch that represents less than 1 per cent of the harvest the bay supported a century earlier. Meanwhile, the race was still on for bluefin tuna, even though the entire Atlantic Ocean's breeding population had fallen to no more than about 22,000 fish. In November 1993, an American fisherman sold a 300-kilogram bluefin to a fish broker for Tokyo's high-end sushi market. The fish went for $80,000.

It is the same the world over. According to the United Nations' Food and Agriculture Organization (FAO), all seventeen major oceanic fisheries are now being fished at or beyond capacity. Sometimes the problem is habitat loss, sometimes it is pollution, sometimes it is overfishing, and often it is a combination of all these things.

The Aral Sea, which once offered up more than 40,000 tons of fish per year, is now a toxic sink. Water diversions have reduced it to a shadow of its former self, and all twenty-four species of fish that supported commercial fisheries are believed extinct.

In the Caspian Sea, sturgeon harvests are now 1 per cent of their 1950s levels, which is a decline in itself no greater than the collapse of the Fraser River's white sturgeon fisheries, which dropped from a peak catch of more than 1.1 million pounds in 1897 to 33,500 pounds in 1902, continuing to show up occasionally in the following years only as an unintended catch in the Lower Fraser's salmon gillnet fisheries. (In 1994, a spate of mysterious sturgeon mortalities, believed to be linked to deteriorating water quality, prompted a total ban on sturgeon harvests in the Fraser's commercial, recreational and aboriginal fisheries.) The Black Sea is little more than a dump site for the chemical and organic pollutants flushed into the Danube, Dniester and Dnieper rivers, which drain half of Europe, and only five of the thirty species that once supported commercial fisheries in the Black Sea are still in existence, forcing a reduction in the Black Sea fish catch from 700,000 tons to 100,000 tons between 1983 and 1993.

On the U.S. West Coast, fishing fleets are starving for salmon. Hydroelectric dams, massive overfishing by Canadian and U.S. fleets, the loss of estuarine habitat to urban development, the trashing of spawning streams by the logging industry and poor ocean survival rates all combined to take a cruel toll on salmon from the American side of the 49th parallel. In the face of such tragedies, the White House seems as helpless as any besieged Third World presidential palace, only slightly more useless than Canada's federal government. President Bill Clinton's major contribution to the growing West Coast salmon crisis was to grant disaster-area status to parts of Oregon and Washington. Before the 1994 fishing season began—a season that degenerated into a preposterous fish war between Canada and the U.S., which the White House was powerless to control—the Federal Emergency Management Agency had already offered disaster relief to Clatsop, Columbia, Coos, Douglas, Lane, Lincoln and Tillamook counties in Oregon, along with Clallam, Grays Harbor, Jefferson, Pacific and Wahkiakum counties in Washington State.

On the Columbia River, salmon populations had been severely

depleted by the 1930s, but there were still 16 million fish that returned to the river every year before dams choked the life out of it. By 1994, the Columbia's wild salmon populations had dwindled to about 5 per cent of their former abundance. In 1988, U.S. commercial fishermen caught 1.8 million coho salmon bound for their home streams in Washington, Oregon and California. Stream enumerations suggest that coho from Washington State southwards are not quite extinct as a species, but in 1994 the U.S. Pacific Fishery Management Council recorded the commercial coho catch as zero. By 1995, coho in the entire U.S. West Coast, from the 49th parallel south, was being considered a candidate for protection under the U.S. Endangered Species Act.

There was a time, on Canada's West Coast, when lingcod ranked fourth in value among the species harvested by B.C. commercial fleets, after salmon, herring and pilchards. One of the largest fish off the B.C. coast, the lingcod can grow to twenty kilograms in weight and more than a metre in length. It is an unusually territorial fish, rarely straying from its home reef, and the male jealously defends egg nests tucked into crevices in rocks. Comprising a substantial component of Coast Salish fisheries for thousands of years, the modern commercial lingcod fishery began in the mid-1800s. Primarily a small-boat, hand-line fishery, it was thriving as a commercial enterprise by the late 1860s in the waters off Victoria, throughout the Gulf Islands and into Burrard Inlet. More than two hundred boats fished the Strait of Georgia area. But the rapid expansion of trawl fisheries for bottomfish preceded a decline in hand-line lingcod landings from about 4,000 tons a year in the 1940s to an average of 1,400 tons a year in the 1950s. Stocks continued to decline into the 1980s, and while the trawl fleet moved upcoast to continue fishing, the small-boat hand-line fleet continued in its home waters, shrinking to only about ten boats by the time the Georgia Strait fishery was closed in 1990, and chances that commercial lingcod fisheries will ever open there again are considered slim to none. The recreational fishery, which accounted for more than 80 per cent of the Strait of Georgia lingcod catch during the final years of the commercial fishery, continued, even though the swift and violent tug of a lingcod on the end of a sportsman's jig line had become an increasingly remarkable event.

Abalone fisheries along the entire B.C. coast were also closed in 1990. In a pattern that has become typical throughout the world, rapid industrial expansion of these fisheries far outpaced any scientific understanding about stock size, stock composition or spawning patterns. With prices reaching $40 per pound in its final, pathetic years, the abalone fishery showed landings that dropped from 481 tons in 1977 to about 100 tons in 1980, and in the five years before the fishery was finally shut down, the coastwide quota never exceeded 50 tons. "There's been such a dramatic change in abundance . . . it is mind-boggling," DFO shellfish coordinator Bruce Adkins told *Westcoast Fisherman* magazine in 1995. "Areas where the fishery landed hundreds of thousands of pounds, there is virtually nothing left . . . One of the things that may be limiting the ability for recovery is that density has been reduced to such a low level that successful spawning and recruitment is limited." As the Pacific Stock Assessment Review Committee noted in 1994, "There will probably not for the foreseeable future be a legal wild fishery for any sector." The stocks are so depressed that poaching for the black market may be sufficient to prevent the rebuilding of abalone populations in even the most remote corners of the coast.

The dramatic decline in the abundance of pilchards is every bit as mind-boggling. Pilchards—Pacific sardines, closely related to French sardines—once supported fisheries from B.C. to California. The rapid development of the pilchard fisheries prompted scientists to warn of the necessity of research and controls as early as 1933. John L. Hart, one of the pioneers of fisheries research in Canada, warned about the consequences of unbridled expansion: "There is . . . an urgent need for information as to the extent of the pilchard stock," he wrote, "and the effect which the fishery is likely to have upon it in order that further investment may not be made unadvisedly." But the warnings went unheeded, and within twenty years, the pilchard fishery was over. There is still no consensus about what really happened. The prevailing view in the scientific community is that little-understood ocean phenomena had something to do with the decline. Old-timers on Vancouver Island's west coast say that overfishing played its part, too. All that is certain is that a fishery once third in value among B.C.'s fisheries, behind only salmon and herring, and second in volume among Canada's fisheries, behind only North Atlantic cod, is now less than a dim memory. Most British Columbians have probably never even heard of the fish, and never knew that Quatsino Sound, Kyuquot Sound, Esperanza Inlet, Nootka Sound, Sydney Inlet, Clayoquot Sound and Barkley Sound once bustled with pilchard reduction plants and canneries.

Similar collapses threaten several other fisheries on Canada's West Coast today, from the smallest to the largest.

Pacific herring, which ranks second only to salmon in value to the commercial fisheries of the coast, came a hair's-breadth away from being fished into extinction in the late 1960s. Herring play a vital role in the structure of marine ecosystems as food for various fish, mammals and birds. After their populations rebounded in the early 1970s, a new herring roe fishery began, to meet the demand for a delicacy produced almost exclusively for the Japanese market. By 1995, the roe herring fishery was supporting 150 seine boats and more than 1,000 gillnet boats that land between 30,000 and 45,000 tons, worth up to $80 million annually. While fisheries managers say they are satisfied that harvests are now sustainable and stocks are in generally good shape, long-time coastal residents, native observers and recreational fishers insist that the commercial fishery is removing too much herring, depriving too much food from local chinook stocks. Countless herring spawning areas have been "fished out" and are not recovering. In 1986, federal biologists conducted an extensive survey of commercial fishermen in the Johnstone Strait and Strait of Georgia area and concluded that, in the straits' waters alone, about 170 locations where herring formerly spawned were substantially diminished or barren.

B.C.'s geoduck gold rush through the 1980s routinely produced annual coastwide landings of 5,000 tons, but by 1994, landings were down to less than 2,500 tons. Because geoducks—huge, long-necked clams—can live more than a century, it is not likely that depressed stocks will soon rebuild. Inshore rockfish stocks are in trouble in many areas along the coast, and they exhibit similarly low stock recruitment and slow growth rates (the rockfish taken by commercial and sports-fishers are often fifty years old and older). Sea cucumbers are also in decline throughout the coast, with landings in 1994 about one-third what they were six years earlier.

During at least two periods in the history of the West Coast's commercial fisheries, the great north-coast halibut banks were almost completely depleted. Many of the most productive fishing grounds were actually producing no fish at all by 1915, and some grounds have never recovered. It took years of negotiations between Canada and the U.S. to produce a halibut agreement, but by the early 1970s, the halibut fishery came close to collapse again, this time as a consequence of bottom-dragging by foreign trawlers. The foreign trawlers' halibut bycatch (bycatch is ostensibly the unintended catch of one species during fisheries intended for another species) was curtailed with the extension of the two-hundred-mile zone in 1979, but the recovery was short-lived. Although rules established in 1944 stipulate that trawl-caught halibut must be thrown back over the side, B.C.'s own trawl fleet has routinely taken hundreds of tons of halibut as bycatch, killing juvenile halibut in the process. As a result, halibut stocks are once again declining. The trawl fleet was catching an average of 1,000 tons of halibut every year from the 1980s through the 1990s. With landed halibut prices as high as $5 a kilogram, the waste of halibut in B.C.'s trawl fisheries may well amount to more than 15 per cent of the entire landed value of B.C.'s regulated halibut longline fishery.

B.C.'s trawl catch has grown by staggering proportions over the past twenty years, targeting fish populations about which fisheries scientists know next to nothing. The fleet's landings have increased from less than 30,000 tons in 1977 to annual averages well in excess of 100,000 tons through the 1990s (almost all of the increase can be attributed to a single species—hake). B.C.'s trawl fishery is now not only the largest fishery by volume on B.C.'s coast, it is also probably a fishery most British Columbians have never even heard of.

Overfishing by trawlers has already pushed some stocks to obvious and serious decline, such as rockfish and Pacific ocean perch. Controls are tenuous at best, and in some years, more than half the fish caught by B.C.'s trawlers aren't even landed in Canadian ports. In some years, the bycatch mortalities associated with the trawl fishery may well have exceeded the worst of the planet's most "dirty" fisheries, including the infamous (and now banned) mid-Pacific driftnet fisheries.

And in the Pacific salmon fisheries, we are already fishing the remnants. The salmon populations that by the mid-1990s provided the bread-and-butter fisheries for gillnetters and seiners in Juan de Fuca Strait, Johnstone Strait, the Strait of Georgia and the lower Fraser River—the primary salmon-fishing grounds on Canada's West Coast —are mere zoo populations compared to their preindustrial abundance. Several analyses conducted during the 1990s by biologists Thomas Northcote, Dana Atagi and M. D. Burwash show that in the century following the rise of the canneries in the late 1800s, the Fraser River's pink salmon runs suffered a 5.5-fold decrease; coho salmon populations suffered a 7.7-fold decrease; chum salmon populations declined 2-fold; chinook salmon populations declined 5-fold; and sockeye populations declined 5.1-fold. Since the 1980s, there have been marked increases in some Fraser River sockeye and pink populations, which most scientists attribute to improved ocean survival rates. There have been no corresponding increases in other Fraser salmon populations, however. In the case of Fraser River coho salmon, returns continue to decline precipitously.

At this late date, it is difficult to fully appreciate the enormity of
✗ these losses. One of the few scientists who has attempted to estimate "pre-contact" salmon abundance is the eminent biologist William Ricker, whose early work Northcote has built upon. Ricker's research, undertaken from the 1940s to the 1980s, provides a glimpse of the way things were. From his work, we can begin to imagine what a spectacle it must have been during the "big year" sockeye runs to the Fraser River, when perhaps as many as 160 million sockeye salmon moved through the Strait of Georgia, across the shallow sandheads of the estuary and into the mainstem, and then on through the mountains to their natal creek-fed lakes. The movement of an organic mass of that size across the face of the globe must have represented a colossal ecological event on the continent's landscape, on a scale equal to the migratory convergences of great buffalo herds on North American plains. Nothing of the kind has been seen on the Fraser in living memory: the average annual return of Fraser River sockeye during the late twentieth century has been about 7 million fish.

Still, as comparatively paltry as 7 million sockeye may seem, the 1980s and 1990s were actually good years for Fraser sockeye and pink returns compared to the preceding three decades. Northcote, professor emeritus at the University of British Columbia's Westwater Research Centre, says these modest recoveries cast the Fraser River in contrast to the "disaster scenarios [in] most other regions of the Pacific northwest (except Alaska)."

North of the Fraser—with the possible exception of recent abundances of Skeena River sockeye—salmon species have suffered drastic declines. On Vancouver Island, coho salmon populations have declined by at least half, and chinook salmon populations have declined almost four-fold. Similar rates of decline have been recorded in salmon species on the central coast, the Queen Charlotte Islands and the north coast. On the Skeena River, chinook populations have declined by at least half.

Steelhead salmon throughout the coast continued to suffer a general decline in all major river systems throughout the 1980s and 1990s. By 1993, summer steelhead populations were known to occur in fewer than thirty streams on Vancouver Island, and, on average, those populations have been reduced by about half, edging them close to candidacy for the provincial threatened-species list.

Up and down the B.C. coast, salmon harvest rates of 85 per cent and more are still routine, particularly in the case of south coast coho and chinook populations, and the loss of distinct genetic stocks of salmon continued through the 1990s. As biologist Brian Riddell pointed out in 1993: "In southwestern British Columbia . . . one third of the spawning populations known since the early 1950s have now been lost or decreased to such low numbers that spawners are not consistently monitored."

So while it is true to say that what happened in Newfoundland could not happen here (because there is no single biomass that accounts for the vast majority of B.C.'s fisheries jobs), it is also true to say that it has already happened here, and it is happening here. And if the prevailing fisheries policies on this coast are not soon abandoned, it will likely continue to happen here until it is all gone.

The situation has become so stark that some biologists have concluded that the best use they can make of their skills in the late twentieth century is to gather the genetic material of remnant fish stocks, store it in gene banks and reintroduce the populations during some later, more enlightened age. "It has to be done. I don't care how unfortunate it is," says biologist Brian Harvey of the World Fisheries Trust. "I don't care what anybody says. As we argue about these things, these fish are disappearing. We need to do everything we can, and this is just a tool, just another tool. It just has to be done. I want my kids to be able to catch wild fish when they grow up."

By the mid-1980s, Harvey was working with the International Fisheries Gene Bank, collecting milt from a variety of vulnerable, threatened and endangered species from rivers throughout the planet. He has successfully cryopreserved the genetic material of several populations within more than a dozen species. In South America, where hydroelectric dams have threatened populations of commercially significant species, Harvey has preserved milt from fish such as the cachama, a creature that reaches six to eight kilograms and eats fruit that falls from trees along the banks of rivers in Venezuela, and which has now disappeared from the Magdalena River. In Brazil,

Harvey's gene bank has been working to save the genetic material from populations of salminus, bracyplatystoma and prochilodus—all locally important commercial species—on the San Francisco, Grande and Doce rivers. By the beginning of the 1990s, Harvey was developing a gene bank for vulnerable and threatened populations of British Columbia fish species, including coho, chinook, sockeye and steelhead.

Working with the Shuswap Nation Fisheries Commission, the Carrier-Sekani Tribal Council, B.C. Hydro and other agencies, Harvey has gathered the genetic material of distinct salmon populations on the Thompson River, the Endako River, Lewis and Lemieux creeks, the Coquihalla River and the Puntledge River on Vancouver Island. Only about two dozen steelhead are known to have returned to spawn in the Puntledge in 1994. When a spawning population drops below fifty fish, the population is considered beyond restoration.

"So, while all the arguing and the name-calling is going on, some of these stocks could be saved for research, and for the future," Harvey says, adding that his efforts rest on the hope that at least some habitat will remain after a fish population is considered lost: "What good is a backup disk without a computer to run it on?"

Meanwhile, something is happening deep within the Pacific fisheries' main "computer"—the Pacific Ocean—and nobody knows quite what it is. While the North Atlantic is growing colder, the North Pacific appears to be warming, and fish are behaving in ways that confound the best fisheries scientists.

Populations of *Osmeridae*, for example, which include smelts and oolichans, are exhibiting behaviours that defy everything science knows about them. Oolichans were once ranked fifth among the B.C. coast's commercial fish species, but by 1912 the small silver fish were abandoned by the commercial industry when competition from other smelt fisheries around the world pushed oolichans off the shelves. Apart from those harvested by a small group of commercial oolichan gillnetters on the lower Fraser River, oolichans were left to B.C.'s traditional aboriginal communities, which have harvested them for the rendering of "grease," a highly valued delicacy, since the earliest days of human settlement on the coast. In the decades following 1912, oolichan populations throughout the coast were stable, and they remained a stirring reminder of the spectacular abundance associated with most of the coast's fisheries before industrialization.

Then, in the 1980s, oolichan populations throughout the coast began to decline at an astonishing rate. Because they were mainly "Indian fish" with virtually no commercial value, there were no fisheries scientists scrambling to explain what was happening, or why. Sto:lo fishermen reported that the writhing shoals of oolichans that once migrated far upriver were growing smaller each year, and each year the fish were travelling shorter and shorter distances upstream. By 1990, the Fraser River oolichan catch had fallen to about 20,000 pounds, one-fifteenth of 1960s' catches. On the central coast, the Klinaklini River, which takes its name from the Kwakwala word for oolichan grease, was barren of fish in 1994. The story was the same on other rivers. The possible culprits included huge herds of Pacific white-sided dolphins that were showing up in inlets where they'd never been seen before, new pollock trawl fisheries in Johnstone Strait, logging damage to coastal rivers, and shrimp trawl fisheries that had for years been killing oolichans as unintended bycatch.

The trouble with most of these localized theories was that the problem was so widespread. As far away as the Columbia River, where oolichans are known as Columbia River smelt, the total 1994 catch was 5 per cent of the average, but the nearby Wynoochee River, which had never been known to support oolichan populations, was thick with spawning oolichans. Meanwhile, the oolichans' relatives—long-finned smelt and capelin—were behaving just as strangely. Autumn-spawning capelin were believed to have mysteriously disappeared from waters south of Alaska until a small population was found spawning in Knight, Wakeman and Kingcome inlets—in the springtime. Longfin smelt, which are also autumn spawners, were found migrating up the Fraser to spawn in the spring. "If you're asking me what's going on," says Doug Hay, a research scientist at the federal government's Pacific Biological Station in Nanaimo, "the answer is, I don't know, and I don't think anybody does."

At the same time, in the North Pacific, an increasing percentage of the salmon that spawn in southern rivers—mainly coho and chinook —just aren't surviving their sojourns at sea.

In 1993, fisheries scientists Richard Beamish and D. R. Boullion conducted an exhaustive study of salmon catches throughout the Pacific over the past seventy years. They compared the catches of Washington, Oregon, California, British Columbia, Alaska, Russia and Japan to the abundance of copepods (a primary salmon food and a key zooplankton species) and the Aleutian Low Pressure Index. Super-imposed on each other, the graphs illustrate the relationship between catch, feed and climate. The relationship is so closely linked and so dramatic that it would appear to suggest an unavoidable conclusion: little-understood planetary cycles play a much more significant role in the long-term health of salmon than any human-designed enhancement efforts might contribute; twentieth-century attempts to modify the ancient life cycles of salmon populations by building hatcheries, conducting lake fertilization programs or using other technological innovations will probably have little effect, in the long term, on salmon abundance.

An even more dramatic illustration of the relationship between fisheries productivity and climatic cycles can be found in the relationship between the catch of albacore tuna off the southern California coast and the annual width of conifer tree rings along the Pacific Coast, tracked from the late 1930s to the early 1960s. Illustrated as lines on a graph, the annual catch and ring width appear as almost exactly the same line.

There is much that human beings can do, and have done, to disrupt ancient planetary cycles. Industrial development around the world has begun to upset the primary ecological rhythms of the Earth itself, and the emission of greenhouse gases has caused global temperatures to rise so sharply that the eleven warmest years since 1854 have all occurred in the past fifteen years. As the planet's fishing fleets grow ever larger and more catch-efficient, the most disastrous consequences can result from simply being unaware of longer-term, cyclical phenomena that affect fish populations, to say nothing of human-caused impacts on these phenomena.

Mysterious events underway in the North Pacific may be linked to these little-understood natural cycles, or to global warming, or to both.

One such event involves sockeye salmon bound for the Fraser River. Down through the millennia, these fish appear to have travelled the "southern route" almost exclusively. Evidence of this migration pattern can be found in the archaeological and cultural records of aboriginal peoples from throughout the Juan de Fuca, Puget Sound and southern Strait of Georgia area. But in recent years, salmon have been returning to the Fraser River more frequently and at greater rates of diversion via the "northern route," down Johnstone Strait from colder, feed-rich waters to the north. Anomalous behaviour patterns such as these can throw off run-strength estimation systems, wreak havoc with fishing plans and produce massive overfishing, which is precisely what occurred with the fabled Adams River sockeye run in 1994. For all our governments' expertise, the Adams River sockeye staggered back to its home streambeds in 1994 in numbers lower than any year on that cycle year since 1938.

Once the homeward-swimming salmon enter the Fraser River, low water levels and high water temperatures are making the final freshwater ascent more difficult. Throughout the 1980s and 1990s, salmon appear to have been dying in the tens of thousands—perhaps the hundreds of thousands—between the mouth of the river and their natal streams. While a certain degree of mortality during their upriver migrations is to be expected, and is probably exacerbated by the increased numbers of tribal nets in the river set by new, inexperienced fishers, studies by the Westwater Research Centre suggest that the deterioration of the Fraser River is causing dramatically high rates of mortality.

A 1993 study, in which sockeye were captured, radio-tagged and released below the Fraser Canyon, showed that a high percentage of them never make it through the increasingly violent environment of the canyon. Water removals for agricultural use and hydroelectric storage—including the Aluminum Company of Canada's increased reduction of flows into a prime Fraser tributary, the Nechako River—have produced low water levels and high water temperatures that sometimes far exceed the sockeye's ability to survive. While many commercial fishers are quick to blame native "poaching" operations for spawning-escapement shortfalls, Scott Hinch of the Westwater Research Centre says Westwater's findings showed that as many fish were dying of exhaustion as they tried to swim upstream as were caught in all the tribal fisheries of the Fraser basin upstream of the canyon. Five of thirty-two tagged fish were caught in tribal nets, but at least another five sockeye that entered the Fraser Canyon never made it through, succumbing to the increasingly shallow rapids where water temperatures in the back eddies ranged between 17.5° C and 18.8° C, a temperature range known to be associated with high sockeye mortality.

The following year, the fishing season was again dominated by an uproar over "missing fish," allegations of native poaching and a federal inquiry into the management of Fraser River sockeye headed up by former fisheries minister John Fraser. "Poaching" was indeed a problem, Fraser's study concluded. But that summer, the Westwater

Research Centre's studies produced results that were far more disturbing than their 1993 findings. Of twenty sockeye radio-tagged below the canyon in late July, not one survived the upstream struggle through the canyon. River temperatures were 17.7° C—higher for that time of the season than in any year since 1942—and Westwater's Hinch said as many as half a million sockeye (a number that about equals all of the fish that went "missing" in the Fraser River in 1994) may have died in the increasingly deteriorated river environment.

If present patterns continue, water temperatures will more frequently hover above the thermal preference of all Fraser salmon, particularly sockeye. A 1994 study by Environment Canada's atmospheric environment service examined the impact of global warming on the Fraser basin's salmon populations and concluded that the coming generations of Fraser River salmon, until at least the middle of the twenty-first century, will face altered aquatic temperatures, precipitation-related changes in flow regimes, alterations in river hydrology, increased winter runoff, reduced summer runoff, and increased competition with human priorities for a share of the water itself: "As the human population and societal water requirements both increase, interactions between different competing water resources users in the Fraser River watershed will undoubtedly intensify in the near future. The overall effect of climate change will be to hasten the intensification of interactions which appear to be inevitable in the long-term. It will be crucial to improve water management practises so that sensible tradeoffs between competing water uses (e.g., fisheries, agriculture, forestry, domestic water supply, power generation, etc.) can be made."

Throughout British Columbia, many of these trade-offs have already been made, with or without public consent, for years. Usually, it is as simple as a spawning stream being traded for a gravel pit. Every year, thousands of homeward-swimming salmon are finding that the streams where they were born just aren't there any more, or that the streams have been so degraded there is little point in spawning there. That is what happened to Byrne Creek, and it continues to happen throughout British Columbia's Lower Mainland, where human population growth in recent years has outstripped every other urban area in North America. In its 1992 *State of the Environment Report for the Lower Fraser Basin*, Environment Canada concludes: "Concerning habitat, two-thirds of the 96 streams surveyed in the Lower Fraser basin in 1985 were identified as suffering negative impacts from activities such as forestry, agriculture, urban development, hydro development and transportation. The resulting impacts, including water withdrawal, blockages, water contamination, water pumping, and removal of streamside vegetation, are making the supply and quality of habitat an increasingly limiting factor in sustaining fish populations."

The situation is much the same everywhere on the coast, even in the more "remote" watersheds, where destruction of fish habitat is commonplace. Erosion caused by clearcut logging, along with chemical spills and stream blockages, have all taken their toll. A 1992 survey of selected forest cut blocks on Vancouver Island, sponsored by the B.C. government, found that logging has had some adverse effect on 34 per cent of the fifty-three surveyed streams. Impacts judged as major (23 per cent) and moderate (17 per cent) were found mainly in streams with the highest potential to support fish, including salmon species.

Federal fisheries biologists estimate that three out of every four B.C. salmon that survive to adulthood are caught by commercial, recreational and aboriginal fishers on their homeward migrations. About 90 per cent go to the 4,500-boat commercial fleet, which routinely catches 80,000 tons of salmon annually (about 10 per cent of the world supply of wild salmon), with a landed value of nearly $170 million. The rest of B.C.'s salmon catch is taken by the recreational and tribal fisheries.

More salmon have been caught commercially in B.C. waters during the past ten years than in any comparable period over the past forty years. But the salmon are smaller than they used to be. In 1994, American fisheries biologists John H. Helle and Brian Bigler surveyed forty-seven separate runs of pink, sockeye, chum, coho and chinook salmon throughout the North Pacific. Of these populations, forty-five showed a marked decline in the average individual weight of the fish. In nine of the populations studied, the average size of individual fish had dropped by as much as 25 per cent over the past twenty years. It is not clear whether ocean conditions alone, or ocean conditions combined with a kind of weeding out of bigger fish by commercial fishing and changed conditions in coastal river systems, hold the clues that could explain the phenomenon. In any case, smaller fish have a harder time making their homeward migration. They build inferior nests, their eggs are smaller and the emerging fry are less hardy.

Globally, the world's total annual fish catch has risen from about 2 million tons in 1850 to about 55 million tons by the end of the 1960s, approaching 100 million tons by 1990. But there is a limit to what the ocean can produce. U.S. fisheries historian James McGoodwin warned in 1990 that while the world's oceans may be capable of yielding as much as 100 million tons of fisheries products annually, that kind of a catch could not be sustained without dramatically altering fishing patterns. Without such interventions in the fishing methods and catch rates of the world's fleets, "the world's catch may soon begin to decline." He didn't have to wait long to see whether he was right. After reaching 100 million tons, the world's fisheries production immediately faltered, fluctuating between 97 and 99 million tons between 1990 and 1994.

Even if significant restraints are placed on the world's fishing fleets, major shifts in management strategies are undertaken and world fish production stabilizes at a sustainable level—itself a tremendously tall order—fish is likely to become increasingly scarce and expensive throughout the coming decades as a consequence of increases in the world's human population. The world is hungry. It is variously estimated that fish and other types of seafood account for between 13 and 16 per cent of the animal protein consumed by people on the planet, but it is also true that more than half the world's human population depends on fish for most, or all, of its animal protein. In Asia, more than one billion people rely on fish. Even if the nations of the world succeed in implementing the ambitious "world population plan" that the United Nations' member states adopted at the Cairo conference in September 1994, the world's population is projected to continue to rise to a point somewhere between eight billion and ten billion. And, according to a 1995 Worldwatch Institute report, there will be increasingly fewer fish for people to eat for at least another half-century.

The world is a big place, but it is also as small as Byrne Creek. It is as small as the Coquitlam River, which once supported thousands of salmon spawners. The Stave River was the same. So were the Nicomekl River, Still Creek and dozens of other creeks throughout the Lower Mainland that are now little more than storm drains underneath shopping malls. China Creek, which boasted four tributaries with several kilometres of prime spawning gravel, was once writhing with shoals of coho and chum salmon from a point around Knight Street and 45th Avenue in Vancouver all the way down to its estuary at False Creek. But False Creek was filled in and China Creek was covered over by pavement and channelled through culverts. Musqueam Creek, which empties into the Fraser at the Musqueam reserve, is the only stream left in Vancouver where salmon still come home to spawn.

Musqueam Creek was to Terry Slack's boyhood what Byrne Creek was to mine. And it has not been pleasant for Slack, a third-generation Fraser River gillnetter, to watch Musqueam Creek die such a long and painful death as he walks its banks in the wintertime, pulls debris from its shallows, takes water temperatures throughout the year and records sightings of coho fry that become less frequent with every year. Cutthroat Creek, a tributary of Musqueam Creek, is the last cutthroat stream in Vancouver, and on a June afternoon in 1995 I found myself there with Terry Slack as he prepared for his annual ritual of scooping up cutthroat fingerlings in tin cans from stranded pools to release them in brighter water downstream.

In the cool of the salal and cedar, with the roar of traffic on Marine Drive reduced to whispers through the trees, Slack walked the creek, approaching each pool slowly and quietly so as not to scare the cutthroat fry back underneath fallen branches from the places they gather, wherever water dribbles in, bringing new oxygen. But there are none here.

Maybe they made it further down, Slack said. We walked further along the creekbed, but we found no fingerlings. Instead, there are three labrador retrievers splashing and rolling in the shallow puddles, off the trail, directly below a sign that reads: "Cutthroat Creek. This tiny creek is home to an endangered strain of cutthroat trout. Cyclists and dogs cutting through the channel are damaging creek life and drainage. Protect fish habitat."

The owner of the dogs identified herself as Hilary. She said her dogs were doing no harm. She wanted to know what role these fish play in the environment, and she asked how they get out of the creek, and whether anyone catches these fish, and what possible use they could be. She said, "I consider myself an environmentalist," and she said her dogs were doing what raccoons do, and it was "natural." Slack tried to explain. This is the last cutthroat creek in Vancouver, he said. She said: "So?"

Slack wandered away into the bush and sat down on the banks of a stream that isn't there any more. After a few minutes, he was thinking out loud about his plans for the fall. "We can push the regional district to pump well water into the creek," he said. There are all these little things to get around to, all these little things that can be done, and the conversation turned from the little things to the big things and back again, as these conversations inevitably do.

"These things have got to be done," Slack said. "Look. I'm a fisherman. I know that we are going to have to change our methods. There's lots of things we can do. Right now, the way we're fishing, it's a hopeless situation. The way we're treating these fish is a tragedy. We're going to have to change the way we think. All of us. I mean real, serious change."

That is how the conversation goes. It is much, much later than most people seem to think, but it is not yet too late.

Human communities are tremendously resilient, and so are fish populations. The great Horsefly sockeye run, which had been reduced to a mere 1,100 spawners when Slack was a boy, back in the 1940s, had replenished itself to more than 10 million fish by 1989. Throughout the world, there are fisheries that work, and they hold some answers for all of us. On Canada's West Coast, there are fisheries that have clearly worked well, some for thousands of years, some for less than a decade. Salmon populations throughout the Fraser basin have shown themselves capable of remarkable recoveries, so much so that Thomas Northcote concluded from his painstaking research: "Habitat, fish and fisheries of the Fraser basin can be sustainable. Whether they will depends on development of appropriate social and institutional changes which must accompany our fisheries management . . . into the twenty-first century."

And as the twentieth century draws to a close, there are more and more people like Slack out there, showing the world what they can do. They are working with near-extinct populations like the coho and the cutthroat trout of Musqueam Creek and Cutthroat Creek. They are working on salmon creeks throughout the coast. They're working on problems of overfishing and habitat loss, and they're building common ground among the most intractable, combative players within a variety of fisheries. They're working on solutions to some of the most difficult fisheries management questions of our age.

For Slack, despite all the declines he has seen, there is reason to be hopeful: "There are a lot of little lights at the end of the tunnel. Fishermen will change, everybody will change if you show them what can be done, if you show them what it could be like in the future."

But it will mean, as Slack says, real, serious change.

Real, serious change.

CHAPTER 11

⚓

A Purse of $630
and a Gold Cane

*A diversity of cultures adapted to local environmental conditions
has largely been replaced by a world culture whose defining characteristic
is its extremely rapid consumption of resources. Artisanal fisheries have
been replaced by industrial fisheries, community responsibility by
government responsibility, economic self-sufficiency by economic
interdependence, and sustainable use by kinds and intensity
of use that are environmentally damaging.*

—Elliott Norse, *Global Marine Biological Diversity:
A Strategy for Building Conservation into Decision Making*

WE KNOW LITTLE ABOUT HIM, ALL THESE YEARS LATER. WE KNOW HIS name was Hans Helgeson. We know he was a federal fisheries officer who was trusted with particularly difficult assignments. We can surmise that he must have been a favourite of the coastal fishing companies, because when he retired in 1910, the canners honoured him with a banquet and a letter of appreciation for the part he played in the destruction of the Babine, Gitksan and Wet'suwet'en fisheries.

Helgeson played his small part in the history of British Columbia's fishing industry during a time when the West Coast's salmon-canning monopoly was extending its operations to the farthest corners of the British Empire, British Columbia's northwest coast. While the industrialization of the salmon fisheries at the mouth of the Skeena River had begun as early as the 1870s, it wasn't until the early years of the twentieth century, when the great Fraser River salmon runs were already suffering from overfishing and showing signs of dramatic decline, that the push to the North began in earnest. By 1900, the prevailing intelligence in the industry was that Skeena River salmon would be the next big thing.

By 1905, the B.C. Packers Association (precursor to B.C. Packers Ltd., which dominates the commercial fishery on much of Canada's West Coast to this day) had already established its Balmoral, Cunningham and Standard canneries at the mouth of the Skeena. The canneries relied almost exclusively on harvests of salmon headed for their natal streams throughout the Skeena watershed, which includes the Bulkley River, the Kispiox and the Babine. Several other companies had set up canneries on the coast and on the islands in and around the Skeena's broad, sweeping estuary. The ABC Packing Company had built its British American and North Pacific canneries. There were J. H. Todd's Inverness cannery, the Wallace Brothers' Claxton cannery, and Carlisle, Oceanic, Cassiar, Skeena River, Alexandria and Village Island. It was like some kind of spectacular, steam-and-iron gold rush.

The canneries established themselves within pre-existing fisheries maintained by the north coast's tribal communities down through the millennia. In the course of the canneries' transformation of tribal fishing patterns, there were the usual skirmishes, standoffs, petitions, outrages, controversies and compromises. The canneries came to the northwest coast at a time when disease and dislocation had depopulated dozens of once-powerful native communities and reduced them to pauperism. For much of the tribal nobility, these were days of unimaginable shame. The tribal commoners were one moment in the thrall of prophets and messianic cults, the next moment awestruck by the new spirit medicine doled out by Salvation Army captains, Methodist preachers and Anglican priests. When the industrial revolution appeared out of nowhere in their midst, with all its managers and clerks and ear-splitting machinery, most native communities were resigned to making the best of it. Most aboriginal leaders tried to adapt to the changing world. And most of the canneries relied heavily on aboriginal communities for labour. Few canneries would have survived their first season—to say nothing of the fishing seasons for decades to come—without aboriginal fishers supplying the canning lines.

Far upriver, however, there were still the holdouts. They were mainly Gitksan people, who were satisfied to make the best of both worlds, and Wet'suwet'en people, who preferred their own fishing stations, where they had maintained elaborate systems of traps and weirs, and of trade and barter, for thousands of years. And there were the Babines, upriver cousins of the Wet'suwet'ens, who were quite content to continue in their traditional fisheries. The problem was that the upriver people were catching the same fish the canners wanted. And the canneries were always short of labourers. Something had to give.

Helgeson's assignment on the Upper Skeena was preceded by the standard public controversy about Indian overfishing, Indian poaching and other alleged Indian depredations on or near the spawning grounds. The tribal fisheries had to be brought to an end. Publicly, the canners warned that the canning industry on the Skeena would be "destroyed" if the upriver tribal fisheries were allowed to continue but were adamant that the upriver natives would not suffer from the loss of their fisheries. Privately, the canners estimated that the number of canneries at the Skeena mouth would double if the upriver fisheries were shut down and the prohibition against native people selling their fish was enforced. As Geoff Meggs points out in *Salmon: The Decline of the British Columbia Fishery*, the canners also threatened to withdraw their support from the Liberal Party unless action was taken.

Federal Fisheries Minister Louis Brodeur demanded at least some semblance of evidence that an assault on the tribal fisheries was justified. Brodeur turned to Hans Helgeson's boss, John T. Williams, inspector of fisheries, Dominion of Canada, District 2, for some answers. Williams, another great friend of the canners, put the problem this way: "The trouble is the Indians are so lazy and idle they will not do anything at all ... Let them come down to the cannery and work as all other Indians do, not loaf. The Babine Indians must realize that they must work as the other Indians do, they cannot be spared."

And so Helgeson was sent up the Skeena River in the autumn of 1905. His first destination was the Skeena's far headwaters, the Babine River, where the Babine people maintained their ancient trap-and-weir fisheries (the canneries referred to them as "barricades"), harvesting the salmon that the canners coveted more than any other Skeena-bound fish: Babine Lake sockeye.

Apart from the early traders of the Hudson's Bay Company, some overland adventurers on their way to the Klondike and the occasional troupe of straggling prospectors, few Europeans had ever ventured into

Babine country. Helgeson filed an extensive report describing what he encountered. His report remains one of the few detailed accounts of the type of run-specific, terminal (end-of-migration) and selective fisheries so many salmon biologists in the late twentieth century would be arguing for as an alternative to the destructive mixed-stock fisheries management that came to prevail throughout the coast. Helgeson's account ranks with that of explorer Alexander MacKenzie, who witnessed a similar fishing technology, and a similar fisheries-management prescription, in operation on the Bella Coola River in 1793. What Helgeson found was certainly not a bunch of lazy and idle Indians, as is evident from his report, contained in the annual report of the Department of Marine and Fisheries for 1905:

> The banks of the Babine River have a lovely appearance at this place and a most wonderful sight met our eyes when we beheld the immense array of dried salmon. On either side, there were no less than 16 houses 30 × 27 × 8 feet filled with salmon from the top down so low that one had to stoop to get into them and also an immense quantity of racks, filled up outside. If the latter had stood close together they would have covered acres and acres of ground, and though it was impossible to form an estimate, we judged it to be nearly three-quarters of a million fish at those two barricades, all killed before they had spawned, and though the whole tribe had been working for six weeks and a half it was a wonder how so much salmon could be massed together in that time.

It was precisely this type of fishery that allowed aboriginal peoples to maintain sustainable harvests of salmon on this coast for thousands of years. It is a fisheries management model with a proven track record in ensuring genetic diversity, an example of species-specific, stock-specific and race-specific fishing at a sustainable harvest rate. It is also precisely the opposite of the mixed-stock management that has dominated fisheries on Canada's West Coast for most of the past century.

The irony of it is that Helgeson himself wondered, "When we take into consideration that nearly every salmon stream in the country is barricaded and that this has gone on for years and years, is it not then a great wonder that there are any fish at all left?" In his 1988 study of the history of the Skeena fisheries, Parzival Copes, a fisheries economist with policy development experience in jurisdictions as far-flung as Papua New Guinea and Newfoundland, answered Helgeson's rhetorical question this way: "Perhaps what is the greater wonder is that Helgeson would not show any hint of acknowledgment that the survival of healthy salmon stocks over the thousands of years of Indian fishing with their accustomed devices was obvious proof that the barricades, as operated by the Indians with alternating openings at their accustomed rates of exploitation, were entirely compatible with stock conservation."

Helgeson and his colleagues set about their work at the Babine River. He advised the local chiefs that their traps had been outlawed, and so had their salmon-based economy: if they wanted to fish, they would have to use gillnets like everybody else, and if the Babine people wanted to fish for money, they would have to go down to the coast and fish for the canneries. The barricades had to go.

The standoff on the Babine River was the first in a running series of showdowns between federal fisheries officers and Babine, Gitksan and Wet'suwet'en communities. The controversy came close to insurrection. Shots were fired; there were arrests and riots and fistfights. The Catholic Church intervened on the natives' behalf and there were gentlemanly differences of opinion in Ottawa between the fisheries minister and the Indian Affairs minister. It was as though the players were reading from a script, with many contradictory twists and turns, that was to be handed down through the generations in British Columbia. Some native people opposed the holdouts. Some fisheries officers supported them. There were "solemn agreements" that were broken and forgotten. By 1910, it was all over anyway, and Helgeson was toasted at his retirement dinner by the thankful canners, who gave him a purse of $630 and a gold cane.

In the years since then, the mixed-stock, non-selective fishing techniques favoured by the commercial fishing industry have severely damaged the salmon resources of the Skeena River. Several species have been seriously weakened, and several stocks within those species are all but gone. Exacerbating the problem, the federal Salmonid Enhancement Program has pumped millions of dollars into the construction of artificial spawning channels at Babine Lake to boost the productivity of the canneries' favoured fish, Babine Lake sockeye. When the commercial net fleets (primarily gillnetters) fish heavily on runs of Babine sockeye, other co-migrating runs are harvested at rates far higher than they can sustain, and other Skeena River species—notably chinook, coho and steelhead—have suffered major declines as a consequence. Biological diversity (to say nothing of cultural diversity) is sacrificed. As Parzival Copes observes: "Reduction or destruction of the smaller wild stocks is a serious matter. This is so not only because these stocks will no longer contribute to the total salmon catch. A more important consideration, probably, is the reduction in the available gene pool."

Over the past hundred years or so, the pattern that has unfolded on the Skeena River has repeated itself again and again around the planet. While it may rarely be so dramatic, and perhaps rarely so charged with corruption, it is a pattern that is nonetheless every bit as crippling to human communities, as disastrous for fish populations and as damaging to sustainability.

As often as not, the victims of this pattern are communities bound by a variety of ties (language, localized skills, local knowledge, family, tradition, etc.) to specific geographic areas, specific fishing technologies and specific fish populations. By their very nature, these communities are indigenous to particular fishing grounds and are engaged generally in small-scale, site-specific, usually selective fisheries. ("Indigenous" in this sense does not necessarily mean aboriginal.) Because the health of these communities is bound to the health of specific fish populations, or a suite of fish populations, their fisheries tend to be sustainable. If the community strays from sustainable fishing, the community dies. With the globalization of capital, the open access policies of state-managed fisheries, the rapid development of technologies and the deployment of newer, faster and more catch-effective fishing fleets, indigenous fishing communities throughout the world have ended up victims of their own success.

Elliott Norse, senior scientist at the Center for Marine Conservation in Washington, D.C., describes the phenomenon this way: "Unlike the indigenous people, the technologically superior group has the option of moving on when the region's resources are exhausted and therefore has little incentive to adopt traditions of sustainable, conservative use. It can earn virtually all the cash benefits from exploiting a stock of fishes, a mangrove forest, or coral reef, but almost none of the environmental costs. Instead these costs are paid by the local communities that formerly depended upon these resources for survival and had developed ways of managing them sustainably."

This is precisely the scenario that has unfolded on this coast, in the Babine fisheries, in the Japanese herring fisheries in the Gulf Islands, in the distinct community of commercial salmon trollers around the Strait of Georgia and in the Strait of Georgia's lingcod hand-line fleet, among others. It works like this: A stable community enjoys a long-standing relationship with a specific population of fish. A new fleet of boats appears on the horizon, or a new fishing technology is taken up by a handful of community members who can afford it, or an outside interest "invests" in the production of commodities from the community's traditional preserve, or a new regulatory regime is imposed from afar. This is called progress. Sooner or later, things begin to break down. Old fishing patterns are abandoned. Old skills become redundant. Competition intensifies, new rules replace old customs, and before long the fishing isn't so good any more. Newer, bigger and faster boats roar off to the fishing grounds every season while old boats rot on the beach or sink at their moorings. Soon, the fish are gone.

Of course, it is usually a lot more complicated than that. It is rarely easy to sort out the good guys from the bad guys—often, there are no bad guys and only a few good guys—and it is rarely so simple a matter as the old ways being better than the new. But the rapid decline in the planet's myriad fish populations over the past century is a mirror image of the growth and destructive catching power of the planet's fishing fleets, in the same way that genetic diversity in fish populations has declined at a pace and on a scale that mirrors the decline in the cultural diversity of maritime and fish-dependent human communities.

There is a common and perfectly rational assumption about the role of the fishing industry in this phenomenon, and it goes something like this: "Gee, the industry wouldn't deliberately overfish the stocks. That would be suicidal." A related assumption is that the fishing industry's main problem can be traced to the presence of "too many boats chasing too few fish."

But the world is a lot more complicated than that. The fishing industry would indeed overfish the stocks, and does so routinely. While this is often a consequence of overcapacity, the problem of overcapacity is not strictly about too many boats. Japan's fishers number about 200,000, but they catch twice as many fish as China's 3.8 million fishermen. In the West Coast's salmon fisheries, the short-hand assertion that there are too many boats chasing too few fish is an oversimplification that tends to be wholly misleading, in the same way that it would be inaccurate to describe the structural problems in the

West Coast forest industry as a case of "too many loggers cutting too few trees." The fishing industry's chronic overcapacity has less to do with numbers of boats or numbers of people than it has to do with catch efficiency and the concentration of catching power—which translates into fish, money and political clout—in a handful of companies. It's about too many boats with too much mobility, too much catching power and not enough accountability.

From their earliest appearance on Canada's West Coast, the key players in the commercial fishing industry have regarded fish populations as resources there to be mined like so much ore. "Sustainability" is a desirable state merely in purely economic terms, and only for as long as the exploitation of a particular common-property resource provides returns greater than alternative investment possibilities will provide. As the term "sustainability" entered the public debate about natural resource controversies, the term "sustainable fishing" came to be kicked around in the fishing industry with such enthusiasm that it lost any real meaning. (A recent industry-government study employed the term in the following ways: "sustainable public policy," "sustainable development of new opportunities" and "sustainable leadership and communication.") While to be candid about what is really meant by "sustainability" is fraught with public-relations implications these days, it was downright fashionable in the industry during the early part of this century. The industry's attitude was well illustrated by the views of D. N. McIntyre of the federal Fisheries Advisory Board, offered in defence of the slaughter of sea lions on the coast by the Canadian gunship *Givenchy*, which was responsible for machine-gunning 6,430 sea lions in coastal waters between 1931 and 1937.

McIntyre remarked: "As it was necessary for the buffalo to be obliterated before the vast area of the prairies were available for cattle grazing and farming with the advent of the white settler, so the sea lion must go with increased fishing on the west coast. I have no more sympathy with pleas for its preservation than I have with the protests of the sentimentalists who deplore the destruction of the buffalo. They stood in the way of the progress of civilization and they had to go. Peace to their ashes."

McIntyre's views are little different from the position the contemporary commercial fishing industry often takes with regard to fish species that get in the way of their prime fisheries. The contempt with which many players in the industry regard steelhead, coho and chinook, species that "get in the way" of the coast's sockeye, pink and chum fisheries, is disturbingly similar to that of their predecessors.

In 1992 and 1993, after years of intense pressure from Gitksan leaders, steelhead anglers and provincial government biologists, federal fisheries planners attempted to reorganize fisheries for artificially enhanced sockeye at the mouth of the Skeena River in order to curb the net fleet's lethal interceptions of co-migrating wild stocks of coho, chinook and steelhead. A degree of cooperation between the various players produced an agreement. During the 1995 fishing season, fisheries managers were under similar pressure to reorganize fishing plans for Fraser-bound sockeye salmon, in order to reduce interception of depleted Fraser-bound coho. But by then, the objective of reducing coho and steelhead bycatch on the Skeena had resulted in a curtailment of sockeye fishing opportunities for the commercial fleet, and the leading spokesman for the United Fishermen and Allied Workers Union (UFAWU) protested that revenues of about $60 million were involved. He asked reporters: "Are we going to waste millions of salmon on the Fraser River like we're doing on the Skeena, chasing after a will of the wisp around things like coho and steelhead?"

From the union official's perspective, there is nothing irrational about such a question. Sometimes it makes better corporate sense to liquidate certain forms of biological capital in favour of more profitable resources, and sometimes it makes good investment sense to deplete biological capital rather than to simply live off the interest. In British Columbia, natural resource companies routinely destroy biodiversity—cut too many trees, catch too many fish, etc.—and there is nothing necessarily "irrational" about this behaviour. It is not even a cost of doing business. As Elliott Norse observes: "People who amass benefits by destroying biological diversity and shifting the cost to contemporaries or future generations are behaving rationally from a purely economic point of view."

There is also a popular notion that the problems of fisheries mismanagement are largely a consequence of the "tragedy of the commons," the idea that the very common-property nature of most fisheries resources encourages overfishing by greedy individual fishers who owe nothing to the resource itself. This idea is most often used as an argument to support privatization of the resource, on the grounds that if the common property was "owned" by its harvesters, everybody would behave responsibly, because everybody would have a direct, private-property stake in sustainability. The problem with this idea is that the conversion of common-property resources to private property has never produced any proof of sustainability. In the early 1970s, Colin Clark, a mathematics professor at the University of British Columbia, made a substantial contribution to the discussion of fisheries management problems around the world by constructing a mathematical model for the commercial exploitation of an animal population. Clark observed that "over-exploitation, perhaps even to the point of actual extinction, is a definite possibility under private management of renewable resources."

Over-exploitation of natural resources can also produce results that serve the private interests of both the harvesting sector and the marketing sector of the fishing industry: sometimes, as a particular fish commodity declines in abundance, its value increases.

This is true of seafood products generally. Historically, fish has been the least expensive source of animal protein throughout much of the world. But as fish becomes more scarce, fish prices tend to rise. The world is growing more hungry, human populations are increasing sharply, and the demand for fish has risen dramatically. As recently as 1960, a kilogram of seafood cost about half as much as a kilogram of beef. But over the past decade, the world price of seafood has risen, in real terms, nearly 4 per cent a year. Even with massive increases in the amount of non-prime fish available, the margin between fish prices and beef prices narrowed, then disappeared, and now seafood is often more expensive.

In his research, Colin Clark considered the "preference of the harvesters for present over future revenues" and concluded that "extermination of the entire population may appear as the most attractive policy, even to an individual resource owner." As irrational as it might seem to a disinterested observer, the dollar in the bank a fish immediately represents may produce profits greater than the value the fish represents in the water. The policy might be particularly attractive in the case of a fish species that exhibits slow stock-recruitment rates. Such a deliberate policy—liquidating the biological capital now rather than living off the interest over a long period of time—can make "rational" sense from a corporate point of view. Clark suggested that this phenomenon may have been behind the fate of the Antarctic blue whale fishery.

As it stands, fisheries resource regulation in Canada has rarely been a simple matter of public management of a common property resource, anyway. As historian Dianne Newell points out, rather than comprising a regime in which conservation-oriented public policy is imposed upon the fishing industry on an arms-length basis, "government policies and regulations usually are responses to pressure from industry to reduce competition and frequently are not in the best interests of other user groups, although the process by which this one-sidedness occurs may be gradual." A defining historical feature of the public regulation of West Coast fisheries has been the elimination of the coastal canning monopoly's real and perceived competitors: from small-scale fisheries and small-boat fleets to aboriginal fisheries, seals and sea lions. Following the pattern that has established itself throughout the world's fisheries, Canadian fisheries policy has hastened industrial ends, devising a system of regulations and priorities that has left biological and cultural diversity at the bottom of the list.

This approach to fisheries management is not restricted to the West Coast, or to Canada. The recent history of fisheries around the planet shows a similar trend. Local management regimes are replaced by large-scale industrial, state-sanctioned regimes; small-boat fleets are replaced by big-boat fleets; cultural diversity is replaced by industrial culture; biodiversity is replaced by monoculture. And all the while, the fish catch rises, fish populations collapse, and a primary source of the planet's animal protein is endangered.

From a mere 2 million tons in 1850, the annual world catch of fish had risen to about 55 million tons by the end of the 1960s. The Worldwatch Institute, which has tracked the history of the planet's fish catches, has shown that the world's oceans had given up about 70 million tons by the early 1970s, and by the 1990s, the catch was approaching 100 million tons. By the 1970s, big-boat fleets were fishing mid-ocean grounds for species that were often far down in the oceanic food chain. Heavily subsidized by governments (the United Nations' Food and Agriculture Organization estimates that large-scale commercial fisheries enjoy about $54 billion annually in various public subsidies), the big-boat fleets had developed to a point by the late 1980s in which their catch capacity was about double the amount required for the available fish.

This race to extinction's cliff edge cannot be slowed down by science alone. It is of no use to say, "Fisheries agencies should manage fish and not fishermen," another widely held opinion about the problems associated with fisheries management. Fisheries are human activities. They are industrial activities. We will have to address the dynamics at work within our own specific human societies, not just within fish populations, if we are to have any hope of conserving what remains and restoring, wherever we can, the fish populations we have lost.

To begin to understand the Byzantine complexity of today's fishing industries and the dizzying pace at which the planet's fish production has increased, it is crucial to be aware of the difference in character, technology, wealth, power and sustainability between "small-scale" and "large-scale" fisheries.

Small-scale fisheries still comprise about 94 per cent of all the planet's fishers, between 14 and 20 million people. The total number of people employed as a consequence of these fisheries worldwide may number as many as 100 million, according to FAO estimates, but they now catch only about half of the planet's fish. The remaining 50 per cent of the catch occurs in large-scale and medium-scale fisheries that may involve fewer than 5 million people, with fewer than 1.5 million directly engaged as harvesters. (The expansion of the world's fisheries since the 1940s has done little to improve the lives of working people in the planet's fishing industries—about 100,000 workers have lost their jobs in recent years, not only as a consequence of fisheries collapses caused by overfishing, pollution and habitat loss but through technological change.) About one-third of the large-scale and medium-scale fisheries' catch is never intended for human consumption. It is harvested for commodities like fish oil and fish meal, which is fed to poultry and livestock. Large-scale and medium-scale fisheries account for virtually all of the bycatch mortality associated with the world's fisheries—which amounts to between 10 and 20 per cent of all the fish caught by humans on the planet.

The distinction between small-scale fisheries and large-scale fisheries is not always obvious. A 140-foot factory trawler owned by a U.S. food conglomerate, fishing in the Indian Ocean, is clearly a different sort of enterprise from a weather-beaten dhow owned by a family engaged in a precarious fishery along Africa's east coast. Pelagic trawlers and driftnet ships owned by Taiwanese food conglomerates clearly represent large-scale fisheries. But herring gillnet fisheries, while they may be undertaken from small-scale skiffs, do not represent small-scale fisheries when they are deployed by half a dozen companies in highly mobile fleets of a few hundred vessels.

The size of the boats, and the number of them, will not necessarily define whether a fishery is small-scale or large-scale. A prawn-and-crab trap boat company that deploys three or four modern and well-equipped boats and as many as 3,000 traps in a season, boasting capital assets in excess of a million dollars, is certainly not a large-scale operation by contemporary standards, and it is more likely to share the characteristics of small-scale fisheries if the business is dependent upon localized stocks, or is confined by regulation or by the knowledge of its fishermen to local fishing grounds. This would be particularly true if the business is the consequence of two or three generations of fishing experience, skill, hard work and investment by members of one fishing family.

Some analysts regard the distinction between small-scale and large-scale fisheries as primarily cultural, and Lester Brown of the Worldwatch Institute points out: "Small-scale fishers—who get the least support from governments—form the backbone of community and cultural diversity along the world's coasts." Fisheries historian James McGoodwin suggests that the distinction may be more usefully identified as that between "ecosystem people" and "biosphere people." Ecosystem fishers are not particularly mobile, and they are consequently dependent on the health of particular marine ecosystems, particular fish species and particular suites of resources. Biosphere fishers may depend on these same ecosystems and resources, but this may well be merely a temporary inconvenience. McGoodwin describes the important difference between ecosystem people and biosphere people this way: "The first tend to be aware that if they deplete their main subsistence resources they risk their own ruin; the second, to subscribe to a 'myth of superabundance,' the feeling that there are always other ecosystems and other resources to exploit should the ones they currently favour run short."

There are some general observations that can be made to distinguish between these two fundamentally different kinds of fisheries. Small-scale fisheries tend to be inshore affairs that demand a wide variety of distinct and specific skills and cultural knowledge. They usually depend on specific, localized fish populations and an intimate familiarity with specific fishing grounds. Large-scale fisheries are almost exclusively offshore fisheries that rely heavily on massive capital investment, large-scale industrial technologies rather than cultural knowledge, and generalist, semiskilled wage labour in place of localized and inherited knowledge and custom.

Small-scale fisheries are far more likely to be sustainable. They tend to be rooted in long-standing, sometimes ancient traditions. They are inclined to employ more appropriate technologies that catch fish much more selectively. They tend to define the character of their communities of origin, and these communities tend to have a direct and demonstrable economic, cultural and social stake in sustainability.

The capital-intensive large-scale fisheries are more inclined to deplete resources, and often directly undermine the stability of small-scale fisheries. Large-scale fisheries are unlikely to have any direct stake in any particular fish population or suite of fish populations. The vessels are usually owned by major firms, many of which are integrated companies with investment not only in the harvesting of sea resources but in the processing and retailing of fish products, and the harvest, processing and retailing of other types of food products as well.

The worldwide trend towards the rapid emergence of highly mobile big-boat fleets and the increased concentration of catching power and wealth among fewer and fewer companies and fewer and fewer boats has played itself out on Canada's West Coast in spades, at the expense of abundance, sustainability, small-scale fisheries and community stability.

Most of these big companies have no direct stake in the health of any particular fish population. Some have no binding stake in wild salmon at all, and have in fact invested heavily in salmon farms, which many industry observers regard as a significant threat to wild fish populations. Some of these companies are integrated national and multinational firms engaged in fish harvesting, processing and retailing, and have no particular stake in the economic health of any community of origin: not a fishing village like Lund, not a city like Prince Rupert, and not even a country like Canada. In the West Coast's groundfish fisheries, company boats with Canadian licences have been deployed to harvest huge volumes of fish in Canadian waters and according to quotas established in management regimes funded by Canadian taxpayers, but the fish are routinely offloaded to foreign factory ships or transported directly to Washington State and Alaskan ports where they are processed by American workers in shore plants sometimes owned by these same Canadian companies.

By 1994, there were about 5,900 registered commercial fishing vessels operating on Canada's West Coast. But more than half of the entire West Coast commercial catch was taken by fewer than 140 big-boat vessels. These vessels are trawlers, and most of them are relative newcomers to the industry. B.C.'s trawlers landed more than 100,000 tons of fish in 1994.

In the West Coast's salmon fisheries, overcapacity, unsustainability and inequity are perhaps the most significant defining characteristics of fleet.

Roughly 4,500 of B.C.'s registered commercial fishing vessels carry salmon-fishing licences. (These vessels commonly carry licences in other fisheries, such as roe herring and prawn, as well.) Most of the salmon fleet—about 4,000 vessels—can be described very generally as a small-boat fleet, divided about equally between trollers and gillnetters. Boats are rarely more than about twelve metres in length and are usually operated by one or two persons. Some are "combination boats," meaning they are equipped with both troll gear and gillnet gear.

Trollers are "hook and line" vessels that deploy several hooks at a time from a series of lines attached to weighted cables. Sometimes requiring the labour of three persons, they often fish far from shore and are equipped with special freezers. Their conventional target species have been coho and chinook salmon, although with declines in those populations coastwide, trollers began to take an increasing share of sockeye allocations by the late 1980s.

Gillnetters deploy nets that are suspended from the surface of the water in the path of migrating salmon, much like a curtain. The net "gills" the salmon. Gillnetters tend to fish inshore, usually around the mouths of rivers or even as far up the Fraser River as Mission. B.C.'s gillnet boats were once relatively "local" boats, rarely straying from home waters. But increasingly they are high-powered, highly mobile boats, with semi-planing hulls and "bowpicker" drums. Although the conventional gillnet catch has been sockeye, pink and chum salmon, their unintended bycatch of co-migrating coho, chinook and steelhead salmon had become a significant cause for concern by the early 1990s.

The remaining 12 per cent of the West Coast's salmon-fishing vessels are seiners. Seiners require the labour of up to six crew per vessel. Routinely fifty feet or more in length, they deploy nets to encircle schooling salmon (primarily sockeye, pink and chum), generally dispatching a skiff to close the circle of the net. The bottom of the net is drawn tight, pursed and hauled aboard. Newer seine vessels are often worth $2 million each. They are constructed almost wholly of aluminum and are outfitted with the most high-tech sonar, radar and satellite-fixed navigational aids. Some seiners double as offshore trawlers. Most Canadians would find they look more like spaceships than fish boats, and the bycatch problems they present are particularly disturbing because of their catching power.

In 1993 there were 536 registered salmon seine boats on the West Coast. These few vessels are now allocated fully 50 per cent of the entire total annual allowable catch of salmon on the B.C. coast, with the remaining 50 per cent of the catch spread among the remaining 4,000 vessels. Because seiners are so "efficient," a typical seiner, rushing from openings on the north coast to those at Johnstone Strait and Juan de Fuca Strait, might spend less than two weeks of the year actually fishing.

A closer look at the ownership of the seine fleet shows even greater inequities. Records maintained by the commercial licence unit of the Department of Fisheries and Oceans show that about one-quarter of the entire seine fleet is owned by two companies, the Canadian Fishing Company (Jimmy Pattison Enterprises) and B.C. Packers (Weston Foods). By 1993, of the coast's 536 seine vessels, 47 seiners were owned by the Canadian Fishing Company and 93 seiners by B.C. Packers. What this kind of concentration means in the overall allocation of Fraser River sockeye, for instance, is that in 1993, with the Canadian commercial catch amounting to 13.8 million sockeye, the seiners deployed by B.C. Packers and the Canadian Fishing Company accounted for, in the most conservative estimate, 1.8 million fish. The direct Pattison/Weston share was about double the entire allocation of sockeye to native communities throughout the Fraser basin, which account for about half the Indian bands in B.C.

Apart from their direct ownership of seine vessels, B.C. Packers, the Canadian Fishing Company, Ocean Fisheries Ltd., J. S. McMillan Fisheries Ltd. and the other major firms represented by the Fisheries Council of B.C. also enjoy contractual relationships of one kind or another with hundreds more vessels—seiners, gillnetters and trollers included. For many commercial fishers, these arrangements—outright vessel ownership by a major fishing company, co-ownership, financial backing, lease arrangements, sales and service contracts, and so on— are crucial to the fishers' continued presence in the industry. The same pattern of increased concentration of resources into fewer and fewer

hands has occurred in B.C.'s fish processing industry. B.C. government statistics show that by 1990, while there were 113 salmon processing companies in operation on the B.C. coast, the top four companies
accounted for 60 per cent of industry sales.

What this all adds up to is a situation in which Canadian salmon,
ostensibly a public resource owned by the Crown and "managed" at a
great cost borne by the public treasury, is quite often caught by a
company boat, loaded off to a company packer, landed at a company
cannery and sold through the company's wholesale and retail networks
(Weston's Real Canadian Superstore, Pattison's Save-On-Foods, and
so on) and marketed under company-owned brand names. The fish
never really enters the realm of "common property."

Across Canada, senior federal fisheries officials have tended to
favour large-scale fisheries over localized, small-scale and small-boat
fisheries. The reasons are not necessarily confined to the inordinate
influence that the big fishing companies wield in federal fisheries
policy, although it would be difficult to overstate their influence. Part
of the reason federal policy favours large-scale fisheries is simply that
such fisheries fit the theoretical model senior managers work within:
that of a central-state authority responsible for fisheries management
and conservation, as opposed to a range of community authorities
overseeing more traditional, community-based management regimes.

From the perspective of the senior ranks of the Department of Fisheries and Oceans, large-scale fisheries—although they often require
massive public expenditures on "management," generally far more
than the management of widely dispersed and localized community
fishing systems would—are theoretically easier to manage. There are
fewer boats, so enforcement and fisheries patrol challenges are not
spread out all over the coast; catches are theoretically much easier to
tally and the fisheries are theoretically much more predictable. There
are fewer parties with which to contend, and the fishing company
managers involved in high-level planning functions are likely to come
from the same social class as the senior fisheries managers and
bureaucrats themselves, in the same way that the owners of these
companies are quite likely to move in the same social and political
circles as the fisheries minister.

But from the perspective of a growing number of harried middle-
management fisheries officials, along with their overworked biologists
and technicians, the increasing dominance of highly mobile, catch-

effective fishing fleets engaged in non-selective, mixed-stock fisheries has actually demanded *increased* management complexity. In the real world, the most nightmarish scenarios become the rule rather than the exception.

In B.C., a useful illustration is the management regime that now governs catches of sockeye and pink salmon as the fish migrate home to their spawning beds throughout the Fraser River system. The regime has come under particularly close scrutiny in recent years, as salmon appear to simply go "missing" by the millions. The regime is governed by arrangements between Canada and the United States under the Pacific Salmon Treaty.

The treaty has its slightly pathological aspects. It is governed by a theoretical framework not unlike the "mutual assured deterrence" policy that prevailed between the U.S. and the Soviet Union during the Cold War period. Only in the case of salmon, it's a matter of understanding that increasingly powerful U.S. fishing boats could completely annihilate Fraser-bound salmon as the fish pass through Juan de Fuca Strait, while equally effective Canadian boats could annihilate U.S.-bound chinook and coho as they pass by Vancouver Island's west coast. The underlying policy governing the relationship: We had all better behave ourselves, or else. The salmon treaty broke down more than once between 1992 and 1995, which was also the period in which the Aboriginal Fisheries Strategy was introduced on the Fraser River.

In the 1940s, Canada-U.S. treaty arrangements with respect to Fraser-bound sockeye and pink salmon were fairly straightforward. The U.S. was contributing millions of dollars to fish ladders and enhancement projects aimed at aiding sockeye production throughout the Fraser River, so, very generally, half the catch was theirs, the other half was Canada's. Meanwhile, Canadian fishing boats continued in saltwater fisheries for salmon bound for other U.S. rivers.

After the 1974 Boldt decision in the U.S. (a court-ordered resolution to Washington State's Indian fishing rights wars), fisheries had to be planned to ensure that Washington State tribes, many of which had ancient traditions of harvests on passing Fraser-bound stocks, ended up with half the U.S. share of the catch. By 1985, fishing opportunities in Canadian waters had to be managed to ensure that specific allocation commitments were met with regard to the seine fleet, the gillnet fleet and the in-river "Indian food fishery."

By 1993, things had become even more complex. The joint efforts of Canadian and American representatives on the Pacific Salmon Commission's Fraser River panel had to ensure that their decisions would satisfy specific allocations of fish both within the U.S. treaty tribes and within a Washington State gillnet/seine/reefnet allocation scheme. In Canadian waters, the catch had to be shared according to elaborate formulae dividing the catch between Juan de Fuca seiners and Johnstone Strait seiners, "outside" trollers and "inside" trollers (including a catch-split within the inside troll fleet for Johnstone Strait trollers and Strait of Georgia trollers), a 60/40 split between Fraser River gillnetters and gillnetters fishing outside the Fraser River, specific saltwater tribal allocations, three separate and specific tribal allocations in the mainstem Fraser below Hell's Gate, and several arbitrary tribal allocations above Hell's Gate.

Before any actual numbers of fish are assigned to the allocations expected by the various sectors of the fleet, the Pacific Salmon Commission is expected to know how many fish are coming home in the first place. To determine this, PSC staff busy themselves with various methods of data analysis and computer simulations to arrive at long-term abundance estimates and pre-season forecasts. Then they oversee test fisheries and calculate run strengths from catch-per-unit-of-effort analyses and other tools. In some cases, the numbers produced by a controversial hydroacoustic fish-counting facility at Mission, far up the Fraser River, are used to estimate the numbers of fish from a particular run of salmon that might still be far at sea, making their way towards the Fraser. Sometimes, the commission allows a seine opening in Johnstone Strait—one of the most lethal and "effective" fisheries on North America's West Coast—in order to produce estimates of how many fish were there in the first place, and how many fish might be left for everybody else.

The system of management has become so complex, and the fishing fleets of both countries have become so powerful, that managing the Fraser River salmon runs is almost like trying to put a person on the moon every year. It is a great tribute to Canadian and American fisheries scientists, and to the scientific and technical staff at the Pacific Salmon Commission, that such a complex management regime has survived at all—to say nothing of the remaining salmon populations the system has been designed to "manage." But it is expecting too much to think that the elaborate dynamics at work within the salmon's migratory range—a range that takes in a sizable chunk of the planet's northern hemisphere—can be made to conform to some kind of management system. No team of scientists and managers can accomplish such a thing, and for all the high-sounding talk about the necessary "ecosystem approach" that natural resource management requires, it was only in the 1980s that biologists with expertise in various aspects of these large ecosystems actually started talking to one another at any great length. There are scientists familiar with the picoplankton that juvenile salmon feed upon during their early weeks of life in interior lakes and rivers; there are foresters and biologists with some understanding of the importance of a healthy forest cover in the watersheds enclosing salmon habitat; there are marine-mammal specialists who might have something to say about the killer whale populations that have developed, over some thousands of years, a unique dependency upon Fraser salmon in the saltwater approaches to the Fraser estuary. "There are dozens of us throughout the various bureaucracies, and we never really talk to each other," a senior DFO salmon biologist explained during interviews conducted for this book.

The elaborate methodology involved in forecasting the likely abundance of salmon in any given year is really no more effective than wondering what's out there as the various salmon populations start showing up close to the coast. As the weeks pass into days, there may be one public announcement declaring an optimistic estimate of run strengths, followed by another announcement declaring a dearth of fish, followed by subsequent upgrades and downgrades as the fishing season proceeds. During the course of the salmon fisheries, run-strength estimates and catch data are gathered and analyzed and updated on a weekly basis. These are duly turned over to representatives from both countries, who are responsible for ensuring not only the fulfilment of the elaborate pre-season allocation commitments but also a measure of control to ensure that enough fish survive the gauntlet to arrive at their spawning grounds in sufficient, "sustainable" numbers.

Sometimes, people make mistakes. In 1992, after an uproar about native poaching and "missing fish" on the Fraser River, then Fisheries Minister John Crosbie confessed to about sixty furious aboriginal leaders in a Vancouver hotel meeting room: "Mistakes were made, but they were mistakes of the heart."

If some management scientist makes a mistake along the way, or if one sector catches more fish than it was intended to, fish get "lost." If some computer simulation is out of whack, or fish migrate in an unusual pattern or behave a bit differently than the textbooks say they should behave, fish either appear suddenly in speculative spreadsheets, as though they fell from the sky, or they simply vanish as scientists decide they weren't there to begin with. If the Fraser River is unusually warm and en-route mortalities are high, or native communities refuse to cooperate with enforcement efforts or catch-monitoring efforts, fish go "missing."

Fish went "missing" by the hundreds of thousands in B.C. in 1992 and 1994. In 1993 as well, an extraordinary number of fish were unaccounted for in the Fraser, and initial counts assessed the discrepancy at about 6 million salmon. But in 1993, these fish weren't "missing." They were "extra." There were fewer frightening headlines about that mistake. But missing or extra, fish numbers continue to be upgraded and downgraded as the weeks and months pass. Missing or extra, it's always the government's fault, or the Indians' fault, or the logging company's fault, or the seals' fault, or anybody else's fault but our own.

"Fishermen are demanding far more than the biologists can deliver," veteran fisheries biologist Carl Walters argued in his controversial 1995 report to the David Suzuki Foundation, *Fish on the Line: The Future of Pacific Fisheries.* Walters wrote: "Somehow, a grave misunderstanding has developed about how good a job the agencies can do at regulating harvests to achieve allocation and escapement goals."

While it would be far too simplistic to cast large-scale fisheries in the "bad guy" role in all this, as though sustainability would be assured by their absence, it is nonetheless true that the issue of whether a particular vessel or group of vessels will participate in the Fraser River fishery, as opposed to fisheries elsewhere on the coast, is often determined in boardrooms in downtown Vancouver and Seattle. Sometimes it's a matter of waking up alone in the below-decks cabin of a gillnet boat anchored somewhere near Rivers Inlet, and deciding whether or not to head south by flipping a coin. But more often an array of complicated considerations are involved, and while there is still that ragtag and eccentric community of gillnet fishers who fish only the lower reaches of the Fraser River, like their parents and grandparents before them, more and more of the commercial vessels involved in the harvest of Fraser-bound fish are highly mobile vessels from anywhere on the coast, or high-tech boats that come from dozens of Canadian and U.S. ports, owned by companies with interests on both sides of the border as well as in fisheries overseas. While the Department of Fisheries and Oceans in the mid-1990s was fast-tracking plans for a type of "area licencing" system, their objective was not to change the fundamental nature of the coast's fisheries. There is nothing "local" or "ecosystem-based" about it. If the Fraser River fails, there is always the Nass, or the Skeena, or Bristol Bay in Alaska, or fish farming for that matter.

Large-scale fisheries have certain needs that tend to overshadow the needs of small-scale, local fisheries, in the same way that capital has basic needs that tend to overshadow the needs of labour. Capital-intensive fisheries demand to be fed huge volumes of fish in order to survive, and the West Coast's salmon fisheries are hardly unique in this respect.

Overcapitalization throughout the world's fleets has reached the point, according to FAO fisheries analyst Chris Newton, where "we could go back to the 1970 fleet size and we would be no worse off—we'd catch the same amount of fish." European Union member states report that the catching capacity of their fleets is on average 40 per cent higher than the amount of fish that can be safely harvested every year. By the late 1980s, Nova Scotia's trawl fleet was estimated to have four times the capacity necessary to harvest annual cod and bottomfish quotas.

On Canada's West Coast, overcapacity in the salmon fleet alone has become so preposterous that it is now almost impossible to quantify it, and technological change is occurring so rapidly that it is almost impossible for fisheries managers to assess how many fish a particular section of the fleet will catch in any given opening. In 1994, it took several scientists several months just to determine how many Fraser-bound sockeye were caught by a few hundred seine boats and gillnet boats during a few brief openings in the Johnstone Strait area. When the analysis was completed, the Pacific Salmon Commission concluded that the fleet had actually caught about one million more sockeye than the commission's technical staff had initially estimated.

Many Canadians are probably familiar with the consequences of fleet overcapacity, thanks to the annual media frenzy surrounding the roe herring fishery. Every spring, television news crews buzz around in helicopters over Baynes Sound and other fishing grounds to assemble videotape of the crush of seine boats and gillnet skiffs as fishers count the seconds to openings that often last only a matter of minutes. The six o'clock news presents the usual report about the madcap herring fishery, featuring on-camera chatter about bonanzas, winners and losers, the whims of Lady Luck and all the rest. What the public rarely sees is the result of this kind of overkill, set out in dreary and unphotogenic columns and tables comparing the roe herring fishery's quotas and catch. In 1983, the coastwide quota in the roe herring fishery was exceeded by 30 per cent. In 1987, the quota was exceeded by 27 per cent; in 1990, the quota was exceeded by 22 per cent, and in 1991 by 25 per cent. To be fair, fisheries managers have done a remarkable job in containing the catching power of the fleet, but the fact remains that in most years—as DFO's own catch statistics clearly show—the roe herring catch far exceeds the roe herring quota.

The federal government has made various attempts to come to terms with overcapacity in the fishing fleets, usually with disastrous results. The most spectacular effort was mounted by Fisheries Minister Jack Davis, who ushered in a multifaceted vessel "buy-back" and licence restructuring scheme that began in 1969 and came to be known as the "Davis Plan." The plan shed several hundred salmon-fishing vessels from the fleet, but the end result was a fleet with much greater catching power (the opposite of the intended effect). Most of the boats lost to the fishery were small boats with a history of small landings, and central-coast native communities were hit hardest. Elsewhere, fishing licences were doubled up on new boats, and the Davis Plan prompted a frenzy of boat-building on the Lower Fraser and the Lower Skeena. Several vessels bought back with Canadian tax dollars were auctioned off, without their licences, at fire-sale prices, and some vessels were bought by American fishermen who licensed them in Washington and used the boats to catch Fraser-bound salmon in American waters.

The cause of fleet reduction has long been advanced by the Fisheries Council of B.C. (the central association of the coast's major fishing companies), which understandably delights in the prospect of vastly reduced boat costs and fewer skippers and deckhands to contend with. (The fishers' union, the United Fishermen and Allied Workers, traditionally opposed fleet reduction, but in later years joined its employers in lobbying for such schemes.) In 1992, a "voluntary licence retirement program" was initiated provincially as a sort of companion piece to the federal government's Aboriginal Fisheries Strategy. The aim of the program was ostensibly to cushion the impact on the industry of increased allocation to constitutionally protected tribal fisheries. But the result, not unlike that of the Davis Plan, is that the commercial catch rose, despite a drop in commercial salmon licences, while the tribal catches since 1992 remained relatively constant. After the 1994 "missing fish" fiasco, senior fisheries planners were instructed by a succession of fisheries ministers to mount another campaign in the cause of fleet reduction (it has such public appeal, after all). But even with a massive reduction in the number of boats, the catching power of the fleet is expected to remain largely undented.

In many fisheries, the only way to ensure a return on investment—*any* return—is to catch a far greater share of the overall quota than the share that would otherwise be the average catch per boat, given the number of vessels in the fishery. This dilemma tends to encourage vessel owners to invest even more capital in their boats and their fleets, in order to compete more effectively. Having invested so heavily, the industry has turned to sophisticated lobbying efforts to protect its increasing investment. In Ottawa, the Fishing Vessel Owners Association of British Columbia employs the same lobbyist as MacMillan Bloedel Ltd.: Nanci Woods of the Capital Hill Group.

The increased fishing efficiency produced by overcapitalization results in increased public expenditures for fisheries management. Management regimes must become far more elaborate and far more precise in order to come to terms with the increased catching power—and lobbying power—of the fleet. Managers find themselves requiring far more detailed scientific information in order to plan fishery openings or justify fishing closures, and management regimes require more costly forms of monitoring, patrol and enforcement. One small mistake in fisheries planning can produce the gravest consequences. Shorter fishing seasons are imposed as the allowable catch is reached faster with every technological innovation. Shorter seasons and more confined fishing areas change fishing patterns, and new fishing patterns invite new technologies and other fish-catching and labour-saving innovations, which in turn require more capital investment.

The consequences of this treadmill are massive private investment, massive public investment, diminishing social benefits, increased rates of error, and often the collapse of entire fisheries.

Historically, the world's big-boat fleets have responded to this kind of situation in a perfectly rational way, from a corporate point of view. Sometimes, fishing companies escape the treadmill by cutting their losses (scrapping their boats, selling them off, or better yet getting governments to buy the boats in fleet reduction schemes) and investing in other endeavours. Just as often, they keep fishing until there aren't enough fish to justify the expense, and then move on to less-depleted fish populations.

Small-boat fishing communities enjoy no such luxury, of course, and nowhere is that fact clearer than in Newfoundland's outport communities. While tens of thousands of inshore fishermen, big-boat deckhands and shore-plant workers were left without work and without hope after the collapse of the northern cod stocks, the companies that had ended up owning much of the catch capacity and much of the processing sector were not left destitute. National Sea Products and Fisheries Products International suffered losses that can be quantified on balance sheets, so they sorted things out and soldiered on the way smart companies do, moving on to other fish populations in other oceans.

CHAPTER III

⚓

Out There in the Deep

There is a simple yet amazingly meaningful Chinese proverb
by Confucius that should be understood by everyone involved in the
Pacific groundfish trawl fishery: "If you don't change direction
you will end up where you are headed."

—Bruce Turris, "The Groundfish Trawl Fishery"

YOU WADE THROUGH THE SHALLOWS, PULLING A BAGLIKE, NETTLE-
fibre net through the water behind you. You gather the net up now and
then to empty the tiny fish it has captured into a reed-woven basket.

That's trawling, an ancient form of fishing that has taken on
innumerable forms throughout the world and throughout the ages.

Among a variety of salmon-fishing technologies anthropologist
Gordon Mohs has documented in the traditional Sto:lo fisheries, trawl
nets were pulled by fishers from paired canoes. Trawling was a practice
employed by the Sto:lo people in the Fraser River from the earliest
times, as far upstream as the turbulent waters immediately below the
Fraser Canyon. On the other side of the planet, the New Testament
fishers who chose to become "fishers of men" 2,000 years ago had
practised a form of fishing on the Sea of Galilee that was, roughly
speaking, trawling. The act of trawling in today's large-scale fisheries
remains at its essence the same as it has always been—you pull a
baglike net through the water, hauling it up now and then to empty
out the fish.

Certain forms of trawling—particularly "bottom-dragging," in
which the net is hauled along the sea floor to scoop up bottom-
dwelling fish—have been cause for concern for centuries. In 1376, the

English House of Commons petitioned the king in a protest about the likely damage that fishing boats dragging trawl nets were causing to sea-floor vegetation. Imagine the kind of damage involved now in the Bering Sea trawl fishery, where ships require tons of steel cable to tow their nets, and some nets are big enough to hold a dozen 747s.

In 1973, a mere 54 vessels on the B.C. coast (less than 1 per cent of the West Coast's commercial fishing fleet) were engaged in trawl fisheries, and only 46 of those vessels were actually classified as trawlers. In 1974, 18 more vessels showed up in the Department of Fisheries and Oceans' records. While this represents an increase in the number of trawlers of less than 50 per cent, the fleet's catch capacity doubled—half of the newly registered boats were new vessels, built with the help of public boat-building subsidies. In the years following 1973, the number of trawl licences issued for B.C. waters remained constant at 142. During the 1970s, however, a substantial portion of the trawl-licenced vessels weren't trawl fishing at all. (Two-thirds of the vessels that carry trawl licences also carry other licences and engage in other fishing activities, such as salmon seining, herring seining and fish packing.) But by the 1990s, all but a small number of trawl licences issued by DFO were being actively fished, often for more than six months of the year, and most of the older vessels had been replaced by newer, bigger boats fitted with the latest in equipment and gear.

By the mid-1990s, trawlers still comprised only about 2 per cent of the West Coast's 5,900-boat commercial fishing fleet. But the trawl fleet accounted for well more than half of all the fish caught by all the boats on Canada's West Coast.

By any standard one might choose to apply, in any analysis that takes into account the characteristics of small-scale fisheries and large-scale fisheries, there can be little confusion about the category into which most of the B.C. trawl fleet appropriately fits.

These are big boats. Department of Fisheries and Oceans statistics show that the average B.C. trawler is a vessel of 106 gross tons with an average length of 19.6 metres. Many of the more catch-efficient vessels are much larger still. Two dozen vessels exceed 26 metres in length. The *Western Crusader* is 27.43 metres in length. The *Arctic Harvester* is 44.81 metres in length. The *Ocean Selector*, the trawl fleet's biggest boat, is 47.8 metres in length.

These are company boats. About one-third of the active trawl fleet is owned by seven companies. These companies are firms with substantial investments in various fisheries and various sectors of different fisheries; some are member companies of large, integrated food conglomerates. The Canadian Fishing Company (Jimmy Pattison's firm) owns two trawlers. Radil Brothers Fishing owns four trawlers. B.C. Packers Ltd. (part of the Weston Foods conglomerate) owns seven trawlers. Ocean Fisheries Ltd. (with holdings in a variety of West Coast fleets and several shore plants) owns eight trawlers. The Ritchie Fishing Company owns nine trawlers. These few boats comprise about one-quarter of the trawl fleet, but they are among the largest boats in the fleet, accounting for a significant share of the overall trawl catch. The *Western Crusader* is a B.C. Packers vessel. The *Arctic Harvester* is owned by J. S. McMillan Fisheries Ltd. The *Ocean Selector* is owned by Ocean Fisheries Ltd.

This is not a community fishery. The trawl fleet maintains only the most tenuous associations with B.C.'s coastal communities. By the late 1980s, the fish processing industries of Washington and Oregon had become reliant on trawl-caught B.C. fish. By the 1990s, the trawl fishery accounted for close to 60 per cent of all the fish caught on Canada's West Coast, but at least 70 per cent of the trawl fleet's catch was either landed directly in U.S. ports, landed in Canada and trucked to U.S. shore plants for processing, or off-loaded directly onto foreign factory ships. Little measurable benefit of any kind has accrued to B.C. coastal communities from the phenomenal growth of the fishery, and many coastal residents are likely unaware of the trawl fishery's existence.

Hake, more than any other species, accounts for the astonishing growth of the B.C. trawl fishery. Also known as whiting, hake is a distant cousin to pollock and Pacific cod. Hake is a fast-growing fish, reaching maturity at three years, weighing about a kilogram on average and reaching lengths of more than a third of a metre. Offshore hake comprises the largest single groundfish biomass off Canada's West Coast. Hake populations spawn in the winter, in the waters off California and Mexico, and then migrate north, passing along the outside of Vancouver Island. Separate populations of hake within the Strait of Georgia and Puget Sound are also exploited by trawlers, as well as by the region's harbour seal populations, for which hake comprises a significant dietary component.

Canadian trawlers caught 5,000 tons of hake in 1979. In 1994, the trawlers' hake catch was 111,000 tons—more fish than the West

Coast's salmon fleet had landed in any single season for decades. Hake stocks have been subjected to the same kind of allocation disagreements between Canada and the U.S. as those provoked by salmon, but they draw none of the public attention that salmon disputes attract. Canada-U.S. hake standoffs have produced catches of migratory stocks that routinely exceed recommended harvest levels.

By the 1990s, even though Ucluelet was seeing some growth in hake-processing jobs, most of the hake caught by Canadian trawlers was still being taken in a fishery that was, for all intents and purposes, a private one. Known as the "hake consortium," the fishery mainly consists of Canadian vessels catching fish in Canadian waters but offloading directly to foreign factory ships. Two-thirds of B.C.'s trawlers routinely participate in the hake fishery for part of their season. A dozen or more foreign ships may be involved in any given year. Since 1980, the foreign ships involved in the hake consortium fishery have included vessels from the Soviet Union, Greece, South Korea, Japan, China and Poland. Between 1980 and 1991, the volume of hake harvested for these "over-the-side" fisheries grew dramatically, from 12,000 tons to 76,000 tons.

Despite the staggering volume of their catch, there are only about five hundred fishers engaged in trawling throughout Canada's Pacific Coast. While many trawling vessels are clean, perfectly seaworthy and relatively comfortable, the nature of the work can be as bleak as any fishery on the coast. Trawl fishers work in round-the-clock shifts in rough seas, through the fiercest storms, and are commonly at sea for a week or more at a time, with brief shore leave. The fishing season lasts several months, and the fishery requires workers with a capacity for endurance and shipwise conduct that is generally found only among people from communities that have long associations with seafaring culture. A surprising number of the fleet's skippers and crew members are Newfoundlanders, many of them from outport communities, and many of them have been at sea since they were children. Some veterans in the fishery, in particular the émigrés from Newfoundland and the Maritimes, have spent half their lives away from their families, beyond sight of land.

The management regime that ostensibly governs the B.C. trawl fishery is so complex and elaborate it makes the Pacific Salmon Commission's management of Fraser-bound sockeye and pink salmon appear as easy as organizing a lunch engagement between two old friends. But complex as it is, the regime is no match for the biological complexity of the fish that trawlers catch—an array of poorly understood species with a spectacular variety of characteristics that has confounded the best scientists on the coast. These species in turn comprise a variety of distinct populations, and conventional stock assessment methods are almost impossible to apply with any success. Many of these species exhibit unique behaviours, and little is known about their stock recruitment rates. Some of these species are highly migratory or are at least widely dispersed throughout the North Pacific. Some of them include populations that keep to their own unique migratory patterns. Other species are made up of an uncounted number of distinct and separate localized populations, sometimes gathered in communities that appear no larger than extended families, whose members rarely stray from a single, small, sea floor neighbourhood.

While some small sections of the trawl fleet target shrimp and other species, the trawlers' main targets (apart from hake) are placed in the rather arbitrary category known as groundfish. These include rockfish (generally known as "redfish" to Newfoundlanders and Maritimers), pollock, Pacific cod, lingcod, sole (flatfish), dogfish and sablefish (also known as black cod). Non-target trawl catch includes halibut, salmon, herring, crab, oolichan and sturgeon.

Rockfish—mainly member species of the *Sebastes* family—occur in a kaleidoscopic diversity in B.C. waters. Among the main rockfish species harvested and landed by the trawl fleet are Pacific ocean perch, bocaccio rockfish, canary rockfish, darkblotched rockfish, quillback rockfish, redstripe rockfish, rougheye rockfish, sharpchin rockfish, shortraker rockfish, shortspine rockfish, silvergrey rockfish, splitnose rockfish, widow rockfish and yellowmouth rockfish. Other species with little commercial value include blue rockfish, China rockfish, copper rockfish, greenstripe rockfish, red-banded rockfish and tiger rockfish. Rockfish are long-living creatures, and it is only very recently that science has begun to grasp just how long these fish can live, and to consider what their longevity implies for sustainable harvests. A recent survey by the Pacific Biological Station found rockfish off the Queen Charlotte Islands that were more than a century old.

Pacific cod, pollock, lingcod, sole, dogfish and sablefish, Dover sole, English sole, petrale sole (brill) and rock sole are all species harvested throughout the coast by trawlers deploying both "mid-water" trawl gear at a variety of depths and "dragging" gear across a variety of sea-floor types. Just how badly the sea floor has been damaged by bottom-dragging remains a matter of some conjecture. But the anecdotal evidence (primarily from halibut fishers), along with evidence from scientific studies elsewhere in the world's oceans and evidence from the handful of studies undertaken in West Coast waters, suggest that the damage has been substantial. Trawl-caused degradation of the sea-floor environment and extreme damage to benthos and benthic organisms (the sea floor and bottom-dwelling sea life) have prompted localized trawl-fishing bans in jurisdictions as far-flung as Alaska, Australia, Florida, Indonesia, North Carolina and Washington State. Advances in trawl gear, including the development of steel cables in place of ropes and the introduction of heavy drag-chains and tickle-chains to scare up fish into the mouth of trawl nets, have allowed dragging technologies to be deployed in deeper water and on a variety of sea-floor types. The result is increasing alteration and disruption of benthic life and degradation of species composition. Algal beds, sea-grass beds and various "hard-bottom" sea-floor types are particularly sensitive to trawl damage.

Bycatch, one of the most serious environmental problems associated with all fishing technologies, is more common and more extensive in B.C.'s trawl fishery than in any other fishery on the coast, and it is as great a problem here as in almost any fishery on the planet. A 1993 study by the United Nations' Food and Agriculture Organization concluded that trawl bycatch rates are likely greater, in fact, than the bycatch associated with the universally condemned mid-Pacific squid driftnet fishery, which was largely brought to an end in 1992 as a result of international pressure.

Bycatch, otherwise known as incidental catch, is the catch of unwanted, "non-target" species of fish that occurs during the course of a fishery. Usually, bycatch ends up dead (hence the term "bycatch mortality") and thrown back over the side (hence the term "bycatch discards"). Many deep-dwelling fish species rarely survive being caught because their swim-bladders burst during the rapid change in pressure from the sea floor to the surface. Sometimes, fishing regulations actually require fishers to discard certain high-value, non-target fish in order to prevent "high-grading," which is the practice of retaining high-value fish only, tossing back the rest of the catch. Sometimes, trawlers reach their quotas for a certain regulated species early in a trip but continue fishing for other species. Sometimes, dumping fish is the consequence of a skipper choosing the lesser of a number of evils: keep the "overage" and land the excess fish, thereby risking prosecution; "misreport" the catch; or dump the catch.

Fisheries that use species-selective gear, or gear deployed at locations and at times when only the target species of fish is likely to be caught, obviously produce the least bycatch. Such technologies are closely associated with small-scale or "ecosystem-based" fisheries. Weir and trap fisheries are common technologies of this type, and they are typically "clean" fisheries, in that they are generally live-capture fisheries that harvest only the target species. Unwanted fish can almost always be released live.

Weirs—a variety of fencelike structures aimed primarily at migrating stocks of fish in rivers or around river mouths—are generally clean fisheries because of the selectivity they allow. Throughout history, weirs have tended to produce fisheries that ensure only a single population within a single species is harvested. Weirs also provide fishers with the opportunity to brail out unwanted fish and let them continue their migrations. Weirs were perhaps the most widespread aboriginal salmon-fishing technology employed on Canada's West Coast until they were banned by federal authorities under pressure from the coastal salmon-canning industry in the late 1800s.

Trap fisheries are still quite common on the coast, and they are used in the black cod fishery, the prawn fishery and the crab fishery. While Alaskan king crab fisheries are known to produce high rates of bycatch mortality, this is the exception rather than the rule among trap fisheries. Dungeness and rock crab fisheries are not only species-specific: fishers also release female crabs as well as undersized crabs, live, with no mortality.

Although often inappropriate in the salmon fisheries in the way they are used, even seine nets, trolling hooks and gillnets can be quite clean technologies under limited circumstances. Deployed properly, seine nets can allow for the live release of non-target fish. Trolling, which can produce a high bycatch mortality of juvenile fish and non-target fish, can also be relatively selective during certain times and areas of salmon migration. And gillnets, used at specific times, with appropriate mesh sizes and deployed in ways that limit potential bycatch, can also be moderately selective.

On a scale of "clean" and "dirty" fisheries, with selective fisheries that produce no bycatch mortality at one end of the scale and non-selective, high-bycatch, high-mortality fisheries at the other, B.C.'s trawl fisheries fairly fall off the dirty end of the chart.

Most Canadians are only vaguely aware of this country's Pacific trawl fishery. During the course of researching this book, it also became obvious that even B.C.'s commercial fishers have little more than a passing knowledge of the trawl fishery. Federal officials, meanwhile, have not exactly fallen all over themselves bringing the facts of the trawl fishery to light, as an internal 1992 DFO report acknowledges: "Historically, DFO's annual reports have not included [bycatch] discard statistics, even when available. Concern that these statistics could be misinterpreted and/or misconstrued, given the controversial nature of the matter, is understandable."

In a rather grotesque irony, one of the fish species routinely tossed back dead by B.C. trawlers—in astonishing quantities—is turbot, otherwise known on this coast as arrowtooth flounder. In the Atlantic, it was a closely related species of turbot, known to East Coast fishermen as Greenland halibut, that so aroused the sympathies of Fisheries Minister Brian Tobin in 1995 that he staged an international incident to curtail turbot fishing by Spanish trawlers just outside Canada's two-hundred-mile zone. Canadian fisheries patrol vessels boarded and seized the *Estai*, a Spanish trawler; its crew was arrested and taken into custody in Newfoundland. The *Estai*'s net was shipped to New York and unveiled to journalists and shocked onlookers.

While the treatment meted out by Canadian fisheries officials to these two turbot species may seem uneven and inconsistent, both policies make perfect business sense. In the North Atlantic, the Spanish were protesting a quota system that favoured Canadian ships, and Canada wanted to drive the Spaniards off the Grand Banks in order to secure raw supplies for its own badly weakened East Coast processing industry. In the North Pacific, dumping turbot can mean the difference between returning to port with valuable fish in the hold, or coming back with a hold full of highly perishable fish that fetch as little as nine cents a pound.

In a 1991 report filed with the B.C. Aquaculture Research and Development Council, scientists G. R. Silver and Douglas L. Macleod concluded: "A conservative estimate of maximum sustained yield for arrowtooth [turbot] in Hecate Strait, based on the 1987 (survey) figures, is 3,400 to 4,000 tons annually ... It is estimated that 3,000 to 5,000 tons of arrowtooth are dumped at sea annually in Hecate Strait and Queen Charlotte Sound." That is enough turbot to provide a meal for every man, woman and child in Newfoundland, Nova Scotia, New Brunswick and Prince Edward Island. It is more fish than the 3,400-ton turbot quota the North Atlantic Fisheries Organization established for European fleets for 1995, which the Spaniards opposed so passionately. It also represents only a small fraction of the bycatch produced by B.C.'s trawl fisheries every year.

Just as fisheries collapses have followed in the wake of trawl-fishing technologies in the Atlantic Ocean, fisheries collapses have followed trawl fisheries throughout the West Coast, from the most southerly fishing grounds to the most northerly. These collapses have been well documented in studies published by the Fisheries Research Board of Canada, thanks to the efforts of scientists such as J. S. Ketchen, working in the 1940s, C. R. Forrester in the 1960s and S. J. Westrheim in the 1980s. To their great credit, researchers for Greenpeace Canada documented much of this tragic history in 1994 and 1995.

From the earliest days of the commercial trawl fishery on the West Coast, sole was a much-sought catch, and the fishery began in waters close to the major local markets, around the Strait of Georgia. English sole was once an important trawl-fishery catch, but by the 1940s trawl fishing had already caused extreme damage to English sole populations throughout the Strait of Georgia and Puget Sound. There were once substantial populations of English sole around Point Grey, Sand Heads (off the mouth of the Fraser River), Galiano Island, Boat Harbour, Baynes Sound and Salt Spring Island. Severe trawling restrictions were imposed in the late 1940s, but the rules came too late, and these sole populations have never recovered. By the 1990s, English sole trawl landings were coming almost exclusively from the waters of Hecate Strait, between the Queen Charlotte Islands and B.C.'s northern mainland coast.

By the 1960s, the trawl fleet had already fished the coast's butter sole populations to the brink of extinction, and butter sole are still considered "commercially extinct." More than one million pounds of butter sole were harvested annually from the waters around the Queen Charlotte Islands between 1956 and 1965, almost exclusively for animal food, but stocks rapidly declined until the fishery was closed.

Beginning in the early 1950s, high catches of petrale sole were followed by marked catch declines in the waters off Vancouver Island's west coast. Petrale sole were formerly fished heavily by trawlers off Port

Renfrew, on Vancouver Island's southwest coast, but the area was producing only occasional and scattered catches by the late 1980s. By 1960, petrale sole populations had been overfished well into Hecate Strait, and by 1993, scientists with the federal Pacific Stock Assessment Review Committee were recommending that petrale sole quotas throughout the coast be "limited to the lowest possible levels." Much, if not most, of the petrale sole now caught on the coast is caught during the course of trawl fishing for other target species.

By the 1990s, the trawl fleet's rock sole catch appeared to have become mainly an incidental catch during trawl fisheries for Pacific cod, lingcod and English sole in the waters off the north coast and between the northern tip of Vancouver Island and the Queen Charlotte Islands. Some trawlers do target rock sole specifically, but during these fisheries the nets often bring up just as much halibut— a species that is supposed to be allocated exclusively to halibut "longliners" (halibut fishers who set anchored lines with baited hooks along the sea floor). Little is known about what "sustainable" rock sole harvests on B.C.'s coast might be, but in a September 1992 memorandum, Richard Beamish, one of Canada's senior fisheries scientists, warned DFO's groundfish managers that the rock sole quotas established for Hecate Strait were probably 50 per cent "above sustainable."

Trawl catches of Dover sole were minimal until the 1970s, when federal subsidies assisted the fleet in finding Dover sole grounds and fishing at the muddy-bottom grounds these deep-dwelling species prefer, over the continental shelf. Catches of Dover sole doubled between 1988 and 1992.

The fish that end up dead in B.C.'s trawl nets, and thrown back over the side, range from some of the most valuable species on the coast, such as halibut and salmon, to the least valuable, such as arrowtooth flounder.

Trawling for halibut has been outlawed since 1944, but B.C.'s trawlers continue to fish the halibut banks for other species, taking as much as 1,000 tons of halibut every year as bycatch in the process. Trawlers are required by law to throw halibut back over the side (to prevent deliberate high-grading), and at least half of those halibut, according to some estimates, die from the ordeal—producing a waste of at least $2.5 million worth of halibut every year.

Trawling for sablefish (black cod) is also illegal, but black cod too is a frequent trawl bycatch. A study by R. D. Stanley, published in 1985 as a DFO technical report, estimated that the annual bycatch of black cod during trawl fisheries in Hecate Strait and Queen Charlotte Sound alone was adding up to as much as one-third of the entire coastwide quota for black cod.

The bycatch of Dungeness crab is routine, although trawling for this animal is illegal as well. After analyzing results from one of the occasional on-deck observer programs DFO conducts in the trawl fishery, the marine fish division of DFO's science branch concluded that the 1992-93 trawl bycatch of Dungeness crab in Hecate Strait alone, over a twelve-month period, amounted to 400 tons. To put this figure in context, it is important to note that the landed value of this amount of crab would be about $3.2 million. The amount is about two-thirds of the entire total allowable catch of Dungeness crab in the regulated Dungeness crab fisheries of Hecate Strait. It also amounts to about half the entire landed volume of Dungeness crab from B.C.'s northern waters. Luckily for the crab, however, the survival rate associated with crab bycatch is believed to be relatively high, although it is difficult to quantify. Crabs are tremendously hardy. Those mature crabs that survive being scooped up from the sea floor, being hauled up to the water's surface in a trawl net in the crush of tons of fish, and being dumped on the ship's deck, sorted and thrown back, will certainly survive the long, slow descent to the whirling, muddied water of the sea bottom.

Herring are not so lucky. Herring turn up regularly in the reports of Canadian monitors aboard the foreign factory ships involved in hake consortium fisheries, and trawlers have routinely scooped up whole schools of the fish from the earliest days of the trawl fishery on this coast. Captain O. B. Ogmundson, who described "hundreds of tons of herring" dumped in Hecate Strait during the 1950s, "shovelled off the deck and hosed down the side," noted in his 1981 report to the hearings of the Royal Commission on Pacific Fisheries Policy: "There is no way for a trawler to selectively fish as he makes a set ... Herring are squashed, and to suppose that they will live is to refuse to understand the process of dragging."

Massive trawl bycatches of Fraser River green sturgeon over a period of several decades appear to be linked, at least in part, to declines in the sturgeon's abundance.

Little is known to scientists about the green sturgeon, which occurs in only a few rivers on the west coast of North America, sometimes travelling more than 150 kilometres upstream to spawn. Green sturgeon are believed to spend most of their lives in deep waters, and densities of them have been noted in the waters off the west coast of Vancouver Island. For years, green sturgeon have been coming up in the nets of groundfish trawlers, usually unwanted, so they are most often thrown back, with diminished chances of survival. Some trawl skippers and crew members report huge catches of unwanted green sturgeon off Barkley Sound, Kyuquot Sound and Cape Cook. Trawl skippers also report dramatic declines in their green sturgeon catches over the years. Department of Fisheries and Oceans records do cite landings of green sturgeon at coastal shore plants, but the records are wildly inaccurate and unreliable.

University of British Columbia resource management student Jim Echols collected sturgeon data for his master's thesis. "I'd say that green sturgeon are definitely declining on the coast," Echols says. "You see a trend from the 1950s, '60s and '70s, where the recorded landings are ... 4 tons a year, 3 tons a year, records like that. Then in the past ten years [to 1995], nothing." Some trawlers' own catch records show hauls in the late 1980s of as much as 4 tons of green sturgeon, by a single boat, on a single day, in a single haul—but during the same period, DFO records show 4 tons of green sturgeon as the entire catch for the entire fleet for the entire year. A comparable analogy would be a forest company's records showing 2,000 board feet of, say, Rocky Mountain subalpine fir, being pushed through a single sawmill on a single day on a single shift, while the forest ministry's records show 2,000 board feet as the total production of the wood from every mill in the entire province for the entire year. Echols's own observations suggest that as much as 80 per cent of the green sturgeon catches that actually end up being kept and landed at shore plants "didn't even get recorded in the stats ... for every ton that is recorded, I wouldn't be surprised if 10 tons are caught."

Like green sturgeon, oolichan remain largely a mystery to science, particularly when it comes to questions about how they spend their lives away from the rivers where they are known to spawn. Dense shoals of these bright little fish, members of the *Osmeridae* family, arise from several rivers on B.C.'s mainland coast, and they are found in heavy concentrations in the feed-rich waters off Vancouver Island's west coast, where there are no oolichan-producing rivers. Hundreds of thousands of oolichan have been scooped up in the shrimp trawl fishery, which produces the highest rates of bycatch of all B.C's trawl fisheries. (Shrimp trawl fisheries are considered among the dirtiest fisheries on the planet.) "The potential for unwanted catch is very large," a federal research report into trawl-technology improvements concluded about shrimp trawling off Vancouver Island's west coast. "Research surveys of by-catch in 1991 showed 70 per cent by-catch (primarily dogfish and eulachons) by weight. A 1992 research survey resulted in only ten per cent of the catch being shrimp."

Nor is salmon exempt. In American waters, studies that revealed the trawl fishery's salmon bycatch—tens of thousands of salmon, mainly chinook, taken in trawl fisheries off California, Oregon, Washington and Alaska—prompted widespread protests. Unknown to most British Columbians, B.C. hake trawlers fishing in the waters off the mouth of the Strait of Juan de Fuca have routinely taken salmon in their hake nets: 13,517 in 1989, 8,086 in 1990, 7,464 in 1991 and 4,347 in 1992. The counts were taken aboard the foreign factory ships that process the B.C. trawl catch.

On the west coast of Vancouver Island and on the American side of the Strait of Juan de Fuca, chinook populations have been declining at a dramatic rate from the 1970s forward, due to overfishing, poor ocean survival rates and habitat damage. In Washington State, dozens of chinook populations have been extirpated from their natal streams and dozens more are endangered. By 1995, sixty-eight chinook populations on Vancouver Island's west coast were in serious decline. (One of the most productive chinook rivers on the entire south coast is the Somass River at the head of Alberni Inlet, where fisheries managers were counting themselves lucky if 6,000 chinook returned to spawn.) But it is in the very waters where chinook populations are known to be found in their highest density that the largest trawl fishery on the B.C. coast occurs—the hake-consortium fishery. In a count of salmon taken as bycatch during the 1983 hake fishery (8,358 salmon), 75 per cent were chinook.

To date, there have been innumerable reports, several incident investigations and several species-specific and area-specific studies related to the bycatch discards problem in B.C.'s trawl fishery. But the only study that provides a glimpse of the extent of the problem remains R. D. Stanley's study, based on investigations in Queen Charlotte Sound and Hecate Strait over the 1981-82 season.

Among his findings:

- As much as 40 per cent of the fish caught by the B.C. trawl fleet was made up of unwanted species that were thrown over the side. (For almost all finfish discarded, the mortality rate is 100 per cent; for some finfish species with a chance of survival, the mortality rate is at least 50 per cent.)

- The annual bycatch of dogfish, arrowtooth flounder and Pacific cod each amounted to several thousands of tons every year.

- The trawl fleet throws back dead almost as many English sole as it keeps.

- The amount of unreported fish from a particular species that trawlers throw back sometimes equals the species' quota and its maximum sustainable yield.

One of Stanley's more important observations was that as long as the precise, per-species extent of trawl bycatch remained unquantified, fisheries scientists would remain unable to establish reliable sustainable-harvest levels, because estimates about a particular stock's size and its recruitment rate are thrown off by unquantified bycatch and bycatch mortality. Meanwhile, rapid technological advances in the trawling fleet—bigger boats, bigger engines, more sophisticated echo-sounding ("fish-finding") technology and navigational equipment—are making DFO's stock assessment job even more difficult.

Perhaps the most significant tool DFO's biological sciences staff uses to determine sustainable harvest rates is the analysis of CPUE (catch-per-unit-of-effort) upon certain stocks. That tool won't work if the CPUE is constantly changing as catch-efficiency increases, and there have been dramatic increases in trawl catch-efficiency in recent years. Much of that efficiency has been accomplished with government subsidies to the industry from different branches of DFO, such as the Industrial Development Branch and its Fishing Industry Services Branch, from the federal Energy, Mines and Resources' research and development program, and from the B.C. government.

Unlike in the Alaskan trawl fishery and Canada's East Coast trawl fishery, no mandatory, comprehensive observer program has been implemented in B.C.'s trawl fishery. Such measures, including a "complete coverage" program of at-sea observers along with port monitors, would be costly. Just such an initiative had been tentatively proposed by DFO as a user-pay initiative, with charges levied against the industry, but the vessel owners refused to cooperate. By the early 1990s, it was clear to some senior federal fisheries officials that even a partial monitoring program could entirely eliminate the profitability of some sections of the industry; already, many trawl vessels in the overcapitalized fleet were finding themselves forced to overfish and misreport their catches in order to make any profit at all. In a February 24, 1993, internal memorandum obtained under the Access to Information Act, the assistant director of DFO's Biological Services Branch noted: "If the intent of at-sea observers is enforcement of the no-dumping rule, Marine Fish staff are pessimistic that anything less than essentially complete coverage of the fleet would allow a program to meet its objectives. They have several reasons for their scepticism of a partial at-sea enforcement program ... There have been annual overages of rockfish landings compared to quotas, in recent years running 1,300 (1991) and 3,500 (1990) tonnes. Revenues from these overages has become part of the income of the fishery." With observers only occasionally on board, skippers would be obliged to return to port once a quota on a particular species was reached, which would "greatly reduce the value per trip." To make up for lost profits, "industry would have to highgrade even more when an enforcement observer was not on board, negating the enforcement benefits of a partial program."

B.C.'s trawl fishery is caught on the classic treadmill of large-scale industrial fisheries. For far too many boats, the only way to ensure a decent return on investment is to catch a disproportionate share of the total allowable catch, so boat owners invest even more capital in their vessels to increase catching power, even if the investment costs far exceed any immediate increases in fish prices. They push the treadmill around, forcing other vessel owners to do the same just to stay afloat, even if it means hauling nets on the edge of the rules, bending the rules or breaking the rules altogether. While this is obviously irrational from a conservationist's perspective—to say nothing of a plain common-sense perspective—it is perfectly rational business behaviour.

The dilemma was articulated well by Bruce Turris, head of DFO's groundfish management unit, in an internal 1994 discussion paper: "Each and every trawl vessel would like to plug their vessel with fish, but the industry wants to ensure that groundfish are conserved for future generations. Clearly though, if each fisherman met his objective, conservation could not be achieved and the industry's objective would fail ... The groundfish industry wants to provide a consistent flow of high quality fresh fish to a wanting market while minimizing the cost of fishing and associated resource wastage. Trawl fishermen, on the other hand, go out and catch thousands of pounds of groundfish, often discarding as much as they catch, to sell to a buyers' market paying pennies a pound and frequently complaining about quality. But for individual trawl fishermen their behaviour is not only reasonable but often essential. If they don't deliver the fish, someone else will. Meanwhile, debts must be paid, commitments kept, and lifestyles maintained."

Turris goes on to cite "misreporting," "discarding," and "catches ... in excess of conservation targets," adding that "many industry participants are unclear which will go bankrupt first, the resource or the fishermen."

Even in the absence of ongoing and comprehensive monitoring, federal fisheries managers are kept fairly well informed about the fishery's rampant fish dumping, overages, poaching, unreliable monitoring reports, inaccurate reporting and misreporting. The tragic irony in this is that the trawl fishery's excesses are known to senior DFO officials thanks largely to the concerns of many working fishers, shoreworkers and skippers in the fishery itself. In January 1984, the Groundfish Trawl Advisory Committee, comprising the trawlers themselves, identified a long list of chronic problems in the fishery, acknowledging "there is too much discarding and dumping in the fishery," "there is too much misreporting of the landed catch," "there is far too much fishing capacity and effort for the amount of fish available annually for harvesting," and "there is too much bycatch." The evidence has been spelled out for several years in enforcement reports filed with senior fisheries officials on a weekly basis. What follows are some illustrations from these reports, obtained by a Greenpeace Canada researcher under the Access to Information Act.

From the December 7, 1990, weekly groundfish review, prepared by Todd Johannson, Prince Rupert: "Two skippers reported they encountered juvenile fish while fishing at White Rocks. [Name withheld] reported that 30,000 lbs. of juvenile sablefish were discarded on his last trip ... [Name withheld] also reported some sablefish discards along with 40,000 lbs. of small Pacific cod ... he caught all of these fish in one tow.

"... A very angry fisherman (wants to remain anonymous) complained to me that several boats are fishing (and have been fishing) in the closed area in Hecate Strait. He even stated that he heard skippers on the radio commenting what the fishing was like in the closed area. The fact that this fishing was taking place with no apparent enforcement intervention made him wonder why this area is closed. He also went on to say that he can't understand why some short-sighted fishermen would even fish in an area closed to save juvenile fish."

In one 1991 weekly enforcement report, an unidentified fisheries officer complained: "[Names withheld] said that they would like to see Amphitrite Bank closed to trawling while Pacific cod are spawning. They said that a large number of trawlers are continuously hauling their bottom gear back and forth across the Pacific cod spawning areas day in and day out for weeks and they felt strongly there must have been large quantities of Pacific cod spawn that was needlessly damaged."

From another 1991 report: "[Name withheld] made some strong comments about the lack of compliance (by some boats) with the Hecate Strait closure. He does not feel that the present penalties and enforcement methods are deterring boats from poaching, they only keep the honest fishermen honest. [Name withheld] also stated that the recent logbook information being collected from the Straits is only about '60% accurate.'"

From an enforcement report from the Prince Rupert waterfront, for the period January 24 to February 4, 1994: "Two large incidental catches were mentioned by skippers: [Name withheld] recorded 75,000 lbs. of ratfish discards while fishing the Inside Edge, another skipper told me he had a 40,000 lb. tow of Dungeness crab while fishing on the South Flats ... [Name withheld] figures that his *Sebastes flavidus* catch was underestimated by 4,000 lbs. and his *Sebastes brevispinis* catch was overestimated by the same amount."

An undated report, apparently filed in 1994 by DFO's Todd Johannson: "Some processing plant workers from time to time continue to bring to my attention that many trawl vessels are bringing in overages on trip limit species. They say these overages are not readily obvious to DFO personnel when they do spot checks, but they do occur and add up to very significant amounts over a period of time. A few shoreworkers have told me recently that they have a big stake in this fishery too and don't want it damaged. One fellow told me that he would like to either change our fishery plan or else crack down on fishermen and processors for any overages at all."

On July 23, 1993, Johannson reported: "[Name withheld] commented to me that he was unhappy with the Department's lack of enforcement power and presence in the U.S.A. He said that 40% of the groundfish is being landed there and we know little of what is going on. [Name withheld] said that the lack of enforcement in the States is a good incentive for fishermen to land there, instead of Canada."

In response to one Access to Information application, an entire page of an undated "biological sampling" document was blanked out except for a single paragraph. It reads: "He said, 'Contrary to what some people are saying, there are big quantities of ocean perch, greenies and other fish being dumped and it's all because of the current poor management plan that we have and should have changed and improved a long time ago.'"

In July 1993, Johannson reported: "Fishers are worried sick about the trawl groundfish fishery. Many of them were very emotional these past few weeks when they told me that dumping at sea of commercially important species is rampant, and it involves bigger quantities than many DFO people realize, regardless of what anyone else says. They admitted that some excess catches are brought in under the guise of misreporting. Many fishers, however, refuse to misreport, and they dump, and they get sick, and they worry for their future and for the future of the resource."

In his 1994 discussion paper on the future of B.C.'s trawl fishery, DFO's Bruce Turris sums it up well: "There is a simple, yet amazingly meaningful Chinese proverb by Confucius that should be understood by everyone involved in the Pacific groundfish trawl fishery: 'If you don't change direction you will end up where you are headed.'"

In the 1994-95 season, after months of consultation with industry, DFO managed to persuade the trawl fleet to cooperate with a new, experimental management regime that alters the trip-limit and quota system in a way that is intended to reduce fleet costs, simplify management, and reduce dumping and bycatch mortality. The regime established quotas on assemblages or "aggregations" of species rather than on specific species. It remains to be seen whether the new management system will result in any significant change in industry practices or in the trawl fleet's impact on fish stocks: a July 1995 letter from an anonymous trawl crew member does not bode well. Addressed to Fisheries Minister Brian Tobin, the letter was obtained by the David Suzuki Foundation. It describes the crew member's experiences during the first year of the new management regime.

"I recently went dragging," the crew member begins.

"We came in with 120 thousand pounds. To get this we had to go through about 400 to 600 thousand pounds of junk. Junk was all the small stuff and the overs after you caught your quota. Black cod was the worst. Trying to catch 3,000 pounds of black cod we went through 100 thousand pounds of junk, some of it we could keep but most of it went over the side. The amount of small or undersized fish was incredible. As soon as you have caught your quota in one species you start throwing it away on the next tow. We also killed lots of baby halibut and crabs."

This is not fishing, in any conventional sense of the term. This is mining.

At the close of the 1995 season, more new regulations were implemented. The rules established an on-board observer program for most trawlers, which would at least provide better data about bycatch in the fishery. But the rules also provided a legal sanction for bycatch dumping in the form of a per-voyage "bycatch quota" that even the Deep Sea Trawlers' Association agreed was "a disgrace." Eric Wickham of the Black Cod Fishermen's Association predicted "a huge amount of destruction" as a result of the plan, which would allow the trawl fleet to catch and dump up to 800,000 pounds of halibut a year and, theoretically, 100 million pounds of black cod.

It is just like the Chinese proverb says: B.C.'s trawl fishery is headed in the same direction it has been headed in from the earliest days of trawling for English sole around the Strait of Georgia during the 1940s. Pushed along with government subsidies, trawlers continue to drag their way across new fishing grounds. The vessel owners continue to invest in more-faster-better technology, forced farther offshore just to stay afloat. In their wake, trawlers have left dead sea bottom, collapsed fisheries and lost opportunities for the fishers themselves, for shoreworkers and for coastal communities.

Specifically, B.C.'s trawlers are headed for the offshore Pacific seamounts—mountains that rise from the ocean floor—to look for new sources of fish on the seamounts' table-flat mesas. Experimental fisheries began on the Bowie, Union and Cobb seamounts in 1993 and 1995, deploying gear capable of fishing at unprecedented depths. "For the first time," according to a 1995 DFO-industry announcement, "species, stocks and the biological environment in waters up to 800 fathoms deep will be investigated for commercial viability."

By the mid-1990s, B.C.'s trawlers were also headed into the high-seas squid grounds, deep in the Pacific, far beyond Canada's two-hundred-mile zone.

The excesses of the Asian squid driftnet fleet have been well documented. What is not so widely known is that Canada—which played a high-profile role in research and in monitoring the Asian driftnet fleet, assisting the enforcement efforts of the Soviets and the Americans—first became involved in driftnet research in the late 1970s and continued joint research projects with the B.C. trawl fleet into the late 1980s. This research was undertaken not because of DFO's concerns about the environmental implications of the driftnet fishery, but because Canada wanted to get in on the ground floor of the fishery to provide new raw material for B.C.'s overcapitalized trawl fleet. It was only after the Asian driftnet fishery became an international scandal that Canada quickly shifted gears from industrial research to scientific, "conservationist" research.

The high-seas Asian driftnet fishery began because the Asian fishing industry was trapped, like B.C.'s trawlers, in a more-faster-better cycle. By the mid-1970s, Asian driftnet ships had caused the collapse of inshore flying squid stocks, specifically *Todarodes pacificus*. To make matters more difficult for the industry, the fleet's notorious "curtains of death," each up to thirty kilometres in length, were banned in Japan's inshore waters after protests by small-scale Japanese fishers. Searching for new, untapped resources, driftnet ships from Taiwan, Japan and South Korea headed out to the mid-Pacific, fishing primarily for *Ommastrephes bartramii*, otherwise known as neon ("flying") squid.

The mid-Pacific driftnet fishery's interceptions of salmon and steelhead caused alarm in Canada through the 1980s, particularly when it became obvious that the fishery was producing a worldwide black market in driftnet-caught salmon. Saddled with this reputation, the Asian fleet's driftnets, so numerous that tied end to end they could have circled the Earth at the equator, were also found to cause a bycatch of glamorous creatures such as whales, dolphins and seals. Asian driftnet-ting soon came to be regarded as the northern-hemisphere equivalent of the liquidation of tropical rainforests. A 1989 study completed by a team of scientists from DFO's Pacific Biological Station that attempted to calculate the mid-Pacific fishery's annual bycatch mortality came up with these estimates: 50,718 northern right-whale dolphins, 14,825 Pacific white-sided dolphins, 14,045 northern fur seals, 6,245 Dall's porpoises, 780 striped dolphins, a "guess" of 331 tons of Canadian salmon and an estimate of 750,000 seabirds.

Governments, conservationists and fisheries scientists throughout the world worked with great effort to force a United Nations ban on squid driftnet fishing, justified mainly on the grounds of the fishery's staggering bycatch. In response, the Japanese driftnet industry pro-tested that its bycatch rates were lower than those of many trawl fisheries, and that their ships were deploying various devices to mini-mize driftnet bycatch. But Japan eventually complied with the ban, which came into effect in 1992. Taiwanese, Korean and other Asian vessels remained in the fishery, but joint monitoring and enforcement efforts by the United States, Canada and the Soviet Union—aided in part by the interventions of agencies such as Greenpeace and the Sea Shepherd Conservation Society—harried the "pirate" driftnet vessels off the mid-Pacific grounds.

Despite Canada's latter-day conversion to the anti-driftnet cause, Canada's involvement in high-seas squid stocks began in 1979—only a year after Japanese driftnet ships began setting their nets for neon squid in the mid-Pacific. That year, DFO began a series of research fisheries aimed at attempting harvests of neon squid, using both trawl gear and driftnet gear. Also in 1979, in the waters just off the coast of southeast Vancouver Island, a joint Canada-Japan effort deployed a variety of gears in attempts to locate commercial quantities of neon squid. Joint Canada-Japan "exploratory" fisheries continued in B.C. waters in 1980, 1983, 1985, 1986 and 1987: "The use of Japanese expertise and equipment gave Canadian fishermen interested in fishing for the flying squid an opportunity to observe an actual Japanese fishing vessel (the *Tomi Maru*) operating and to learn the fishing methods employed by the Japanese," according to a June 1987 DFO communiqué. The B.C. trawlers involved in the fisheries were the *Ocean Pearl*, the *La Porsche* and the *Simstar*. During the 1983, 1985 and 1986 research seasons, these three vessels fished the same waters, off Vancouver Island's west coast and outside the two-hundred-mile zone off Washington and Oregon, as the *Tomi Maru*.

While the target species was squid, the bycatch mortality produced by these four boats, in three brief "seasons," was enormous. The marine mammals killed included 37 Dall's porpoises, six short-finned pilot whales, four Pacific white-sided dolphins, two harbour porpoises, four northern right-whale dolphins, two killer whales, a Cuvier's beaked whale, a fur seal and two "unidentified" cetaceans. Other bycatch mortality included several hundred salmon and seabirds. At the close of the 1987 season, public reaction was swift and disapproving of the news that forty-four more marine mammals of various species had been killed in "experiments." Canada's brief driftnet fishery was over, and by 1988 DFO's attention had turned to driving the Asians off the mid-Pacific grounds.

After the driftnet controversy gave way to other fisheries controversies, Lee Alverson, former head of the U.S. Bureau of Commercial Fisheries (forerunner to the U.S. National Marine Fisheries Service), and perhaps the leading international expert on fisheries bycatch, published the first comprehensive report on bycatch and bycatch mortality in the planet's fisheries. In his 1994 study, *Bycatch Discards in World Fisheries: Quantities, Impacts and the Philosophical Bases for Their Management*, Alverson concludes that the mid-Pacific Asian driftnet fishery, for all its excesses, was a clean fishery relative to contemporary trawl gears and other technologies.

The authors of an extensive study published in 1994 by *Ocean Development and International Law*, a leading academic journal, reached the same conclusion. "On the basis of a comparison of discard rates and wastage of the driftnet fisheries to those of other fisheries," the study concludes, "something appears amiss in the treatment of the driftnet fisheries. Were they to be held to the same standard as other fisheries and harvesting industries, could the penalties that have been imposed on them be considered rational? Even the highest driftnet fishery bycatch rates were lower than prevailing rates in fisheries harvesting important components of the world's fish and shellfish production." The study's authors—William Burke, Mark Freeburg and Edward Miles—list a series of major fisheries with much higher bycatch rates than those of the squid driftnet fishery, and most of those dirtier fisheries were trawl fisheries. The list includes hake trawling, shrimp trawling, rock sole trawling, cod trawling and deepwater flatfish trawling.

But now that the Asian driftnetters are gone, B.C.'s trawlers, with the help of federal and provincial subsidies, are preparing to take their place on the high seas.

As early as 1988, DFO was funding offshore squid-trawl experiments by the trawler *Howe Bay*, and a major push for a return to the squid grounds began in 1995. Industry groups joined with the B.C. government to draft plans for an "environmentally friendly" squid jigging fishery, and while these discussions were underway back on shore, trawlers were engaged in an experiment on the mid-Pacific squid grounds, using trawl nets so large that they require two trawlers to pull them in paired tows. DFO's Responsible Fishing Operations Division backed the 1995 offshore squid trawling experiments. The Deepsea Trawlers Association provided assurances that a "selective device incorporated into the trawl" will minimize the bycatch mortality of marine mammals and birds.

Said Trawlers Association president Doug March: "B.C. trawlers can expand their operations. New opportunities could be realized and the pressure on traditional inshore stocks will be relieved."

More, faster, better.

CHAPTER IV

⚓

The Triumph of Mechanism

Long time ago, the salmon didn't run up the river.
The coyote and his relatives were hungry. Coyote said, "I don't
know why the fish don't run." He said, "They must do something to
keep the fish from running." He went to the river to find out what was
happening. He jumped in the river, and changed himself into a piece of
wood so he could float. For some time he floated, until he got stuck in
somebody's net. There he was stuck. He was there until two Indian
doctors (who were women) went to check their net.
... The coyote stayed for four days, to find out what the Indian
doctors were up to, and to see how he could beat them. On the fourth
day, early in the morning, when he woke up, he went to the river, broke
all the nets, and that way he broke their power, so they couldn't stop
the salmon from running upriver anymore.

—from the Shuswap story "Coyote Breaks the Dam"

BY THE LATE TWENTIETH CENTURY, THE DAM COYOTE DISMANTLED AT the beginning of the world had been rebuilt over the course of a mere hundred years or so. It had become an almost unimaginably complex edifice. There was the General Agreement on Tariffs and Trade, the North American Free Trade Agreement, the Convention for the Conservation of Anadromous Stocks in the North Pacific Ocean, the Canada-U.S. Salmon Treaty, the Pacific Salmon Commission, the Department of Fisheries and Oceans, the Fisheries Act, the Indian Act, the Forest Act, the Aboriginal Fisheries Strategy, the British Columbia Water Act, the Workers Compensation Act, the Unemployment Insurance Act, the Fraser basin Management Board, the Fraser River Watershed Committee, the Fraser River Advisory Committee and the Fraser River Estuary

Management Program. These are just a few of the components of the statutory and institutional complex through which salmon must somehow make their way, and under which human communities, from the further reaches of the Fraser River watershed to the most isolated coastal communities, must somehow maintain their fisheries.

The salmon management complex is itself only one component of the elaborate management regime that operates under the federal government's constitutional responsibility for fisheries in Canada. The federal fisheries bureaucracy began in 1867 with the Marine and Fisheries Department, and since 1979 it has been known as the Department of Fisheries and Oceans. On the West Coast, the industrial and commercial regulation of the fishing industry is largely the responsibility of the B.C. government, which also regulates freshwater fisheries by authority delegated from the federal government. But overall, the primary responsibility for fisheries management falls to DFO, whose annual budget exceeded $750 million by the 1990s. About one-third of DFO's budget is spent on corporate policy and program support; another third is spent on science; the final third is taken up by operational budgets, inspection services and international affairs. The department's workforce exceeds 6,000 people.

The structure is governed by the federal fisheries minister, a deputy minister, and several assistant deputy ministers and regional directors. DFO is divided into dozens of departments, branches and units, each more or less within five separate program areas: policy and program planning; science; fisheries operations; inspection, enforcement and international affairs; and corporate management. There are dozens of specialized sections within the bureaucracy, including sections devoted to recreational fisheries, habitat management, market development, economic analysis, federal-provincial relations, hydrographic services, marine cartography, biological sciences, fishing industry services, inspection and trade policy, as well as a "ship branch," a "realty and equipment" department, a small craft harbours branch and various special task forces, services and bureaus. Aside from the Fisheries Act, DFO administers more than a dozen other statutes, such as the Atlantic Fisheries Restructuring Act, the Fish Inspection Act, the Fisheries Development Act, the Fisheries Improvement Loans Act and the Fisheries Prices Support Act. The department is also involved in about twenty international bodies, among them the North Pacific Marine Science Organization, the Great Lakes Fisheries Commission, the

Canada-Greenland Joint Commission on Beluga and Narwhal, the United Nations' Food and Agriculture Organization, and the International Union for the Conservation of Nature and Natural Resources.

DFO spends about half its operations budget on the British Columbia coast and employs about 1,500 people in functions that include habitat management on 105 major rivers and the administration of more than 200 harbour facilities, 24 hatcheries and 19 spawning channels. The coast is divided into three main divisions: North Coast, South Coast and Fraser River (which, strange as it seems, is also DFO's operations headquarters for the Yukon). The management structure ostensibly manages and oversees the activities of about 5,900 fishing vessels in a commercial industry that accounts for about 25,000 full-time and part-time jobs in the harvesting and processing of a variety of fish, mainly salmon, herring, groundfish, shellfish and halibut, with an annual landed value often exceeding a billion dollars. In the salmon fisheries, the commercial fishery accounts for about 90 per cent of the catch, with the remainder allocated to the aboriginal fisheries of about two hundred Indian bands and several hundred thousand resident and non-resident anglers.

Below, within and around this edifice are the United Fishermen and Allied Workers Union, the Pacific Trollers Association, the Pacific Gillnetters Association, the Gulf Trollers Association, the Fishing Vessel Owners Association, the Fisheries Council of B.C., the Native Brotherhood, the Aboriginal Fisheries Commission, the Steelhead Society, the Sportsfishing Advisory Board, the Sportsfishing Institute of B.C. and the Fisheries Survival Coalition. And those are just the interest groups in the salmon sector.

At the heart of the value system that props up the increasing complexity and rigidity of these structures and processes are a hubris about our ability to "manage" wild fish populations and a faith in technological "progress" as a remedy for the inevitable damage that such hubris causes. This is the mythology that has come to define twentieth-century approaches to salmon management.

Coyote, whose footprints remain as indentations in boulders at various places in the Fraser River, determined specific access rights to specific populations of salmon among specific Stl'atl'imx and Secwepemc communities. It was Coyote himself who provided weir fishing technologies among the Thompson River Secwepemc. Coyote is said to have instructed the chief of a village in the vicinity of what is now the community of Chase, a short drive east of Kamloops, in the details of fish weir construction; the chief showed his gratitude by offering his daughter to Coyote, who took the young woman as a wife.

Similarly, the contemporary industrial salmon management mythology provides a framework for access rights to salmon, mainly through limited-entry "A" licences issued to commercial fishing boats. The new mythology is as all-encompassing as the old, right down to the deployment of certain specific fishing technologies—gillnets, seine nets and trolling.

Twentieth-century industrial approaches to salmon management owe their origins to the nineteenth-century myth of superabundance, and more importantly, to the myth of salvation through advances in science and technology. This myth was described by one of its most ardent adherents, the American intellectual Timothy Walker (1802-56), in terms as poetic as the words in the old Shuswap story. Walker assigns Mechanism in the place of Coyote. He told the story this way: "Where she [Nature] has denied us rivers, Mechanism has supplied them. Where she has left our planet uncomfortably rough, Mechanism has applied the roller. Where her mountains have been found in the way, Mechanism has boldly levelled or cut through them. Even the ocean, by which she thought to have parted her quarrelsome children, Mechanism has encouraged them to step across."

On Canada's West Coast, the commercial fishery's excesses have demanded an elaborate justificatory edifice that comprises a unique terminology, a shifting set of villains and a fanciful official history.

The Hell's Gate Slide of 1913 has established a central place for itself in this bizarre construction, and industrial histories of the British Columbia fishing industry have lent the event an almost mystical quality. The story has been told and retold. Careless railway builders blasted the side of a mountain into the Fraser Canyon, making it near to impossible for the fish to pass. The disaster of 1913 is generally held to have been the single major cause of the decline of the Fraser's great salmon runs, a catastrophic episode from which the fishing industry has valiantly struggled to recover. What is rarely mentioned is the fact that for all the damage railway blasting caused the great sockeye runs in 1913—and without question, the damage was horrific, particularly for pink salmon—the commercial catch of Fraser River sockeye that same year was a preposterous 32 million fish, a catch that has never since been exceeded or even approached. The closest the commercial fishery has come to pre-1913 catches was its 17.8-million-sockeye catch in 1993 (an estimated 5 million sockeye made it home to spawn), which in itself was about three times the annual average sockeye catch of the preceding forty years.

In a little-known 1987 study, veteran fisheries biologist W. E. Ricker noted that pre-1913 catches of Fraser River sockeye probably amounted to as many as 50 million fish during the "dominant run" years (while sockeye return to the Fraser every year, they are made up of distinct populations of four-year-old fish, with one of those lines usually becoming "dominant" in abundance, partly due to fishing pressure, over the other three lines), and these dominant-run sockeye populations, headed for dozens of spawning homelands throughout the Fraser basin, were severely damaged by the obstructions caused by the Hell's Gate blasts. But the overall decline of the Fraser's sockeye runs had little to do with the Hell's Gate disaster (and as contrary as this may sound, these declines had little to do with the Aluminum Company of Canada's Kemano project, either, despite Kemano's damming of the Nechako River, a major Fraser tributary). As Ricker states: "The initial decline of sockeye of the non-dominant 1903 and 1904 lines, which started at the turn of the century, and the continued scarcity of early and mid-season upriver runs during the 1920s, cannot be ascribed in any significant degree to Hell's Gate or other obstacles. Rather, it was mainly a result of too large a rate of harvest, which is estimated to have been 91 to 94 per cent during 1930-34 and had probably been at least 85 per cent ever since 1900."

But it is of little use to argue for either Coyote or Mechanism as the culture hero most likely to provide fisheries management regimes that might allow sustainable fisheries through the twenty-first century. It is not as simple as that. And regardless of arguments in favour of the restoration of tribal fisheries or for the respect of aboriginal fishing rights, there are other legitimate interests in the Pacific fisheries that must be taken into account. After the few brief days in 1808 when Simon Fraser travelled through the Secwepemc and Stl'atl'imx and Nlaka'pamux and Sto:lo fishing grounds, nothing would ever be the same again. There was smallpox and influenza and tuberculosis; there were the gold rush, the railways, the settlers and residential schools. And then the canneries came. Soon, the old mythology wasn't working any more. The old laws were being broken and forgotten, and the old medicines couldn't cope with the new sickness. Coyote could not break the new power at the mouth of the river.

But by the 1990s, Mechanism had become just as enfeebled. After little more than a century under the new mythology's dominion, only a fifth of the Fraser system's pink salmon runs remained. Only one-seventh of the coho remained. Half of the chum salmon populations were gone, along with four-fifths of the chinook and four-fifths of the sockeye. Meanwhile, the fishing industry had become more strident in its defence of the system established on its behalf, arguing for "industrial solutions" to fix every crack and fissure in the dam. At best, the federal government's management of fisheries is nothing more than crisis management; at worst, it is window-dressing and the allocation of resources to whatever the voters imagine is a good idea at any given time. It now takes up to a decade from the time DFO biologists clearly identify serious conservation problems to the time the department takes any action. Biologist Carl Walters, in his 1995 report *Fish on the Line*, noted: "Chinook salmon collapse in the Georgia Strait was first pre-dicted in about 1979, and we still have not taken the obvious steps needed to reverse that. Newfoundland cod scientists began to recognize that the stock size had been grossly overestimated, and that fishing rates were far too high, as early as 1985; the fishery was finally closed in 1991. Reports warning about coho declines in the Georgia Strait started to appear in 1988; nothing has been done yet in response to those."

Because of the intense jealousies between seiners, trollers and gillnetters in the salmon fleet, the complex allocation systems that have arisen, and the heavy investments tied up in so many boats and so much gear and so much competition on the world salmon markets, fisheries ministers and senior DFO officials often quake at the very thought of taking decisive action to conserve stocks, regardless of the evidence gathered by DFO scientists. To close a fishery invites protests and rallies and unseemly public controversy; to allow a fishery to continue is to gamble that the scientists are wrong and things might turn out all right. A routine approach to such decisions is to avoid them entirely, favouring instead an appearance of action by authoriz-ing major expenditures on costly hatchery projects that have immedi-ate political payoff but potentially harmful downstream side effects.

By 1996, the industry itself has degenerated into utter dysfunction. Technological innovation and overcapacity have produced a ludicrous situation in which only a fraction of the licenced salmon fishing boats on the B.C. coast are really necessary to harvest the available fish, and without continuing the massive injections of federal subsidies the industry has come to rely upon, much of the industry would collapse. The stereotype of B.C.'s "welfare fisherman," or of the rich fisherman who spends his summers collecting unemployment insurance in Mexico, is a gross caricature of the reality for most fishing industry participants, but it is nonetheless true that between 30 and 40 per cent of the coast's salmon fishers could not survive if their sole income was from the fishery, and thousands of fishers and shoreworkers rely almost exclusively on unemployment insurance for most of the year. Less well-known, however—and never the subject of attention-grabbing headlines—is the extent to which the West Coast's major fish-processing firms have come to rely upon federal handouts. According to the audited financial statements of the Fisheries Council of B.C., the West Coast's biggest fishing companies—B.C. Packers Ltd., the Canadian Fishing Company Ltd., J. S. McMillan Fisheries Ltd., Ocean Fisheries and a handful more—are taking in hundreds of thousands of dollars in annual "non-repayable" grants from Canadian taxpayers. In 1991 and 1992, these non-repayable grants included $84,134 in Industry, Science and Technology (IST) funds for research; $206,625 from the IST, $135,000 from Western Economic Diversification funds and $136,000 from the B.C. Trade Development Corporation for export market development; and $65,000 in Western Economic Diversification funds along with $76,400 from the B.C. Trade Development Corporation for "domestic sales promotion."

Meanwhile, DFO's budget was shrinking throughout the 1990s, bleeding funds for such vital programs as the coastal patrolman's association, the eyes and ears of DFO enforcement on the coast, as well as the stream enumeration programs so vital for scientists to develop even the most primitive understanding of what's really going on with the resource. During Fisheries Minister Brian Tobin's tenure, plans were being laid to reduce the 125-year-old federal presence in fisheries and marine regulation to such an extent that the entire federal structure would behave like some retreating army, withdrawing from the coast and heading eastward across the landscape. Lighthouses were closed down. Hatcheries were abandoned. Plans were being laid to merge what remained of DFO with the Coast Guard. Also planned was a major rewrite of the Fisheries Act, which included reducing the 103 pages of sportsfishing regulations to a fifteen-page document, and "partnership" arrangements with fishing industry groups that would allow industry to be involved in decisions regarding limits, number of licences issued and sanctions for fishery violations.

For everyone from the poorest gillnetter to the most senior company official, a virtual seige state exists. This analysis does not arise from the industry's supposed critics or detractors; it is the prevailing view within the industry itself.

In the *1994 Fish Processing Strategic Task Force Report*, an exhaustive study of the commercial fishing industry commissioned by the B.C. government, four hundred leading players in the industry cited:

- A debilitating preoccupation with short-term gain over long-term sustainability, and overlapping government jurisdictions matched only by burgeoning industry sectoral competition.

- Staggering market competition from worldwide salmon-farm production—which almost doubled between 1974 and 1994—causing precipitous declines in salmon prices and destructive market instability. By the early 1990s, Canada had completely lost its European markets to Norwegian farmed salmon production.

- Shore-plant closures and industry layoffs. Despite rising demand for canned salmon, an employment contribution of 25,000 (overwhelmingly part-time) jobs to the B.C. economy and the presence of more than a hundred firms in the salmon-processing industry, the number of major fish processing facilities on the B.C. coast declined from fifty in 1934 to twelve in 1974 and six in 1994, half of them located in the Lower Mainland. The Lower Mainland's three canneries provided 70 per cent of the coast's entire shoreworker payroll, but even that concentration of wages contributed only one-half of 1 per cent of the wages paid in the Lower Mainland as a whole.

- Increasing threats to the survival of salmon populations, despite DFO's annual Pacific budget of $100 million and about $500 million spent to 1994 on the Salmonid Enhancement Program. Competing demands, not the least of which are "untold billions of dollars" foregone in hydroelectric dam construction, weigh heavily on governments.

- Canada's radically diminished ability to make any meaningful decisions about fishing industry policy as a consequence of GATT rulings and the terms of the North American Free Trade Agreement.

- The movement of B.C. processing jobs to lower-wage American ports. Two major B.C. firms had opened salmon canneries in Alaska; other B.C. firms had "dismantled" B.C. processing facilities, moving them to Washington State. While groundfish had come to take up more than half the total volume of fish caught in B.C., only one-third of B.C.'s groundfish were being processed in B.C. shore plants.

Even before the Canada-U.S. Free Trade Agreement (later NAFTA) began to take its toll on the B.C. industry, the southward gallop of fish and capital had become a stampede. By 1990, Steven Spencer of S and S Seafoods in Portland, Oregon, was boasting he could employ "half of Portland" processing B.C. fish if he wanted to, and he was already processing 7,000 tons of B.C. fish every month, mainly because B.C. shoreworkers' wage rates were higher than wages paid in the U.S., where labour unions had been "beat to death." Also in 1990, B.C. Packers closed the groundfish-processing lines at its Imperial plant at Steveston, laid off 150 workers and started sending its boats, with B.C. fish in their holds, to a new B.C. Packers plant, Nelbro Packing of Anacortes, Washington.

The *1994 Fish Processing Task Force Report* proposed a series of recommendations. Most notable in its proposed remedies was a circle-the-wagons "sectoral strategy" against all its challenges; even greater control vested in the fish processing companies; "niche market" research for specialty fish products; securing higher prices for dwindling supplies of fish; government assistance in developing new fisheries from previously "underutilized" fish species; and new sectoral and regional (Canada-U.S.) organizations.

The task force rejected proposals that would dismantle the large-scale fisheries management model in favour of localized, community-driven, small-scale models such as Community Development Quotas. Community quota fisheries were backed, but only for fish species the industry has shown no interest in, and only if no established fishery was displaced. Similarly, proposals for various local salmon management models "cannot seriously be considered a management option" because such models were "insufficiently understood."

The task force recommendations illustrate a profound, almost touching faith in the mythology of Mechanism. But it is a faith that fewer and fewer commercial fishers share, and throughout the industrialized world, its greatest critics, by the late twentieth century, were emerging from its own priestly class: scientists, economists, engineers and biologists.

"Humankind has adopted an arrogant and ultimately self-defeating attitude toward nature that places technological mastery over nature at the forefront of our approach to many environmental problems," writes the University of Georgia's Gary K. Meffe. "This 'techno-arrogance' fails to recognize limitations on, and ramifications of, attempted control of nature."

Writing in the academic journal *Conservation Biology*, Meffe cited "technological applications" to the problem of declines in Pacific salmonid populations as a prime example of "techno-arrogance" gone terribly wrong.

Typical of the failure of technological solutions is the history of salmon hatchery programs on Canada's West Coast. Very generally, a salmon hatchery is a government installation on a salmon-bearing stream that collects eggs and raises them to juvenile fish in confined conditions before releasing them, theoretically allowing higher survival rates and more abundant returns of adults. In reality, the opposite often occurs, along with a range of unanticipated adverse consequences. Fish hatcheries can be used tactically in cases of extreme depletion on small streams and to replenish "fished-out" streams near the stream from which the brood stock has been obtained. Meffe, like most biologists, concedes that under certain limited circumstances, hatcheries can be a useful tool in restoring fish populations. But his analysis of Pacific salmon hatchery programs concludes that salmon populations have continued to decline despite decades of hatchery development. Among his findings: Hatcheries are enormously costly, and they tend to divert government resources away from initiatives aimed at restoring wild salmon populations, making them inherently "unsustainable" because they require continuing, long-term expenditures of energy and capital; hatcheries tend to encourage harvest rates higher than the rates that co-migrating, wild salmon populations can sustain, resulting in declines in the abundance of wild salmon runs that co-migrate with populations of hatchery-reared salmon; hatcheries are "genetically unsound" and "conceal from the public the truth" of real salmon decline.

Elaborate, technology-driven salmonid enhancement initiatives aimed at maintaining or restoring coho and chinook populations on Canada's West Coast have produced alarming failures, some of which have attracted international attention. Typical of these failures have been attempts to pump up declining Georgia Strait coho and chinook stocks. While salmonid enhancement funds have been expended on a variety of projects (hatcheries, spawning channels, lake fertilization programs, and so on) and in varying degrees on all species of West Coast salmon, most of the money has been spent on coho and chinook. About two-thirds of the $500 million spent on the Salmonid Enhancement Program between 1977 and 1994 was spent on chinook and coho rebuilding programs, with no overall improvements in the precarious situation in which these species have been placed. Much of the federal government's salmonid enhancement efforts appear to have done much more harm than good. As Raymond Marsal's report for the World Wildlife Fund shows, the ocean survival rates of fish from almost all hatcheries have steadily declined since the 1970s.

As is the case with the fishing industry itself, SEP's harshest criticisms have come from within, from the most junior engineers to some of the very people who designed and implemented the ambitious federal program.

"We were wrong," says Edgar Birch, who served on the federal salmonid enhancement task force for ten years, from SEP's inception in the late 1970s. "We should have listened to people who were telling us that all these hatcheries were a bad idea. We spent all that money on coho and chinook. We spent all that money on hatcheries. We were going great guns, but hatcheries were a bad idea from the very beginning."

Birch, who also served for five years on the Fraser River panel of the Canada-U.S. Pacific Salmon Commission, has been fishing most of his life. It has been almost fifty years since his first season aboard his gillnet boat, the *Bev-Mark*. Birch fondly remembers the "good years" in the industry. But fishing has become a hard way to make a living, and looking back on all the effort he put into SEP, he spends a lot of his time shaking his head, wondering if there was anything he could have done to change the course the federal government was on.

"They gave us $150 million and they wanted us to produce a return on that, in fish, in five years," Birch says. "[Former Fisheries Minister] Romeo LeBlanc said if we didn't produce those fish in five years, there would be no continuation in funding, so we had to back hatcheries. We had no choice. Our hands were tied. We wanted to clean up the streams for the wild fish, and things like that, but the government said no. They wanted to use the money to employ native people and build hatcheries."

Birch says the hatcheries SEP built produced fish, but only in the short term. Hatcheries pumped out chinook and coho for the tourist boom anticipated in the planning for Expo 86, and the B.C. government spent its own salmonid enhancement funds on glossy brochures to invite Europeans to come to B.C. to catch wild steelhead. Birch and his colleagues on the SEP task force travelled the country, the continent and the world to look at enhancement projects.

Birch, along with United Fishermen and Allied Workers leaders Joy Thorkelson and Bill Procopation, travelled to Sakhalin Island on the Soviet Pacific coast. "They told us we were crazy. They were limiting their hatchery production, and they told us to do the same."

Birch describes the fundamental problem of hatcheries this way: "Salmon aren't like steers or chickens. With hatcheries, you get a lot of fish, but every year they come back weaker."

The Department of Fisheries and Oceans' overall hatchery efforts can't even be justified from a basic cost-benefit perspective. Resource economist Peter Pearse, assigned by the federal government to review fifteen years of effort by the Salmonid Enhancement Program, paints a grim picture of SEP programs and their consequences. Pearse conceded that SEP has made enormous contributions in public involvement, public awareness and public education, and it should be credited with numerous localized success stories. But, Pearse concludes in his 1994 report, wild stocks in some instances may have been "displaced and replaced by enhanced stocks." Overall, "there has been a significant loss in the abundance and diversity of many wild stocks of all species of salmon due to overfishing," and the loss "has been aggravated by the build-up of enhanced stocks which are exploited heavily while wild stocks mingle with them."

It is precisely this co-mingling tendency of salmon runs that requires salmon fisheries management prescriptions and salmon fishing technologies to be specific and selective in order to be sustainable.

But the large-scale management models that govern B.C. salmon fisheries, and the non-selective gears deployed in these fisheries, by their very nature tend to produce unsustainable harvests whenever fisheries are conducted in "mixed-stock" situations: fisheries that harvest several species of salmon at the same time. The management of Fraser River sockeye is a classic example.

Salmon runs exhibit a wide variety of "productivity" rates and can sustain varying degrees of fishing pressure. Some salmon runs have high fry-to-smolt survival rates, and high ocean survival rates, and consequently allow high harvest rates. Other salmon runs have low survival rates and may face particularly difficult obstacles to migration as they make their final ascent to their natal streams, exhibiting high "pre-spawn mortality rates." As a consequence, these low-survival-rate populations can support relatively low rates of fishing. In almost all cases, survival rates and migration difficulties can vary from year to year and from cycle to cycle. Industrial salmon management regimes rarely allow the necessary flexibility or selectivity to account for the many variables that go into determining what a sustainable rate of harvest might be upon any particular run in any given year.

The Pacific Salmon Commission and the Department of Fisheries and Oceans have assigned the dozens of distinct Fraser basin sockeye populations into only four separate "stock" groupings: Early Stuart, Early Summer, Summer and Late sockeye stocks. But within each of these broad stock groupings are several distinct sockeye runs that tend to co-mingle as they make their way home to spawn, before separating to head up different tributaries as they get closer to home. Early in the summer, sockeye bound for Pitt Lake, a few short kilometres upstream from the mouth of the Fraser, tend to co-migrate with sockeye bound for Bowron Lake, which is about as far upriver as sockeye travel in the Fraser basin. Late in the summer (usually late August), sockeye bound for Cultus Lake, only a few kilometres further upriver than Pitt Lake, arrive at the mouth of the Fraser at about the same time as sockeye bound for Adams Lake, hundreds of kilometres within the Thompson River drainage.

Despite its best efforts over the years to determine the relative abundance of sockeye populations within these broad stock groupings in any given year, DFO sets harvest rates mainly on the estimated size of the stock grouping as it approaches the river. While these rates might be perfectly sustainable on the larger populations within each of these particular groupings, an unavoidable consequence of such management techniques is that the smaller, so-called "weak" runs co-mingling with the stronger runs often get badly depleted during the course of what would otherwise appear to be sustainable fishing. A similiar phenomenon occurs in the co-mingling of some Fraser sockeye runs with pink salmon runs, when pink salmon openings produce high rates of harvest on small, "subdominant" sockeye runs. When spawning enumerations are completed, goals established for the big runs may well be met, but four years later, when the offspring of those various runs begin returning to the coast, the big runs can end up returning in numbers relatively larger than the smaller runs within the stock. After succeeding years of mixed-stock fishing on cycle after cycle of salmon populations, the big runs become increasingly dominant, the smaller runs—being "passively managed" (a fisheries management euphemism for "ignored")—begin to disappear, and biodiversity is increasingly diminished.

The kind of gear deployed in the coast's salmon fisheries, and the locations in which these fisheries take place, tend to result in fisheries that are non-selective. Along with trawl gear deployed for groundfish, shrimp and other species, the most "dangerous" non-selective fisheries on the B.C. coast occur in the major fishing grounds for Fraser-bound salmon. These are the gillnet fisheries of the Lower Fraser, Johnstone Strait and Skeena River, as well as the seine fisheries of Johnstone Strait and the Strait of Juan de Fuca. The salmon troll fisheries on Vancouver Island's west coast and the Georgia Strait salmon sports fishery are also notoriously non-selective.

What makes these fisheries particularly dangerous is not just their inability to distinguish between different runs of sockeye. These fisheries are often unable to distinguish between different species of salmon. Sockeye openings for the Juan de Fuca seine fishery have produced a significant bycatch of endangered, non-target chinook and coho runs, for example, just as commercial and native gillnet fisheries for chum salmon in the Fraser River routinely produce dangerously large catches of co-migrating, non-target steelhead.

For years, senior industry leaders and senior DFO officials have been aware that these fisheries management regimes and technologies are unsustainable. In 1988, DFO's own Pacific Region Salmon Stock Management Plan noted that during these mixed-stock fisheries, "some less productive stocks are subject to the exploitation rate applied to the target stock and can be harvested beyond the sustainable levels." The two main salmon fishing areas on B.C.'s south coast are Johnstone Strait and Juan de Fuca Strait, where intensive net fishing for "dominant" runs of pink and sockeye salmon takes place every year. In these fisheries, relatively small runs of sockeye and pink, along with chinook and coho runs, are often intercepted at dangerously high rates. The Salmon Stock Management Plan observed that these relatively smaller salmon runs "cannot withstand the high harvest rates applied to capture the surplus of these dominant runs."

It would not be fair to suggest that DFO has done nothing to address these problems with the coast's major salmon fisheries. But most of the department's attempts have been in the areas of research, monitoring and damage control. As biodiversity is diminished and more and more salmon populations get smaller and smaller, the unintentional interception of salmon from weakened salmon runs grows exponentially more acute. Certainly, non-target, vulnerable and threatened salmon populations are not the only unintended catch of the predominantly non-selective gear types deployed in the coast's fisheries. A 1980 study in the Barkley Sound salmon gillnet fishery found that nets set by a mere seventy gillnet boats were responsible for killing between 175 and 250 marbled murrelets in a single season, an estimated 7.8 per cent of the local murrelet population. But it is the catch of non-target salmon that poses an obvious and direct threat to the sustainability of the salmon fishery itself.

While DFO was pumping hundreds of millions of dollars into hatchery programs to rebuild or at least maintain south coast chinook and coho populations—particularly those in Georgia Strait—the bycatch of coho salmon during sockeye, pink and chum fisheries in Juan de Fuca Strait between 1987 and 1992 averaged 160,000 coho per year. These coho came from scores of streams, many of them badly depleted, on both sides of the Canada-U.S. border. Similarly, between 1987 and 1992, an average of 16,000 chinook were being caught every year as bycatch during seine and gillnet fisheries for sockeye and pink salmon in Johnstone Strait.

Department of Fisheries and Oceans staff have had an extremely difficult time trying to determine what mortality rates might be. Coho mortality rates may be as high as 100 per cent in some fisheries, and chinook mortality rates as high as 80 per cent. Many gillnet fishers do their best to release entangled chinook, coho and steelhead before the fish die, and many seine skippers and crews have gone to extraordinary lengths in their efforts to brail out the unintended catch of coho and chinook in their nets. But the extent of the problem has rarely been quantified, monitoring has been patchy and enforcement is almost impossible.

In the troll fisheries for chinook and coho off Vancouver Island's west coast, the bycatch problem is not confined to a hook's inability to distinguish between a fish from a healthy run and a fish from a weakened or endangered run. A significant bycatch problem in these fisheries is the catch of undersized and juvenile fish. Trollers are required to release the juvenile fish they catch, but the mortality rates for these fish are generally believed to be 50 per cent or more.

"I remember one day when the decks on my boat ran red with the blood of juveniles," recalls David Ellis, a veteran salmon troller and former member of the Commercial Fishing Industry Council, B.C.'s senior fishing industry advisory body. "It was on Swiftsure Bank, near the end of the season. Canada hadn't caught its coho or chinook quota under the [Canada-U.S. salmon] treaty, so they opened up the banks where it's 80 per cent juvenile fish. American fish. Our job was to destroy as many American fish as we could—that's the way I see it. We were all out there, fishing all day, and I was probably killing four or five hundred fish a day. I became a committed environmentalist after that."

Ellis had spent his entire working life in B.C.'s salmon fisheries—as a working fisher, an industry lobbyist and a consultant. But after that late summer day on Swiftsure Bank in 1989, at thirty-eight years of age, Ellis never went back to fishing. It was the "last straw" for him, and he turned his full-time attention to working with small-boat fishers trying to secure a place for themselves in sustainable fisheries, as well as with native groups and conservationists working for changes in fisheries management practices. He has been a thorn in the side of the major industry players ever since.

Ellis makes no apologies for his devotion to fish, which he traces to his youth, fishing for lingcod in the waters around Horseshoe Bay. "I loved those ugly creatures," he says. During his teens, Ellis devised a proposal to restore a depleted chum salmon run in West Vancouver's Nelson Creek, near his boyhood home. His plans never got off the ground, but his commitment to conservation, coupled with his love of fishing, laid the foundation for that character trait that has distinguished so many working salmon fishers on the B.C. coast. Ellis began his life as a commercial salmon troller in the *Minke*, a sixteen-foot hand troller, in Skidegate Inlet on the Queen Charlotte Islands in the early 1970s. It was perilous work, fishing in such a small boat, sometimes dangerously far offshore into Hecate Strait, but Ellis had a young family to take care of, and he was working his way through university, earning degrees in anthropology, science and community planning, at the same time.

After his first years on the *Minke*, Ellis moved up to a thirty-eight-foot troller, the *Fort Winks*, and soon became a "highliner," a top-earning troller. "I killed my share of endangered chinook," Ellis says of his involvement in the industry. "I made a lot of money some years. But looking back on it, we were being used. All of us trollers were being used. The companies wanted our fish. DFO said it was a sustainable fishery, but it wasn't. First, we saw all the coho disappearing out of Hecate Strait. Then it was the same with the rockfish. I was a footsoldier in the salmon wars, and I was betrayed. We were all betrayed.

"We've got to start moving towards more terminal fisheries—fisheries in the rivers, not just on the coast. I never thought I'd say this, but now I think we should probably just shut the salmon fishery down for four years, putting some of the big companies into bankruptcy. We should be looking at some kind of Endangered Species Act protection for these fish, like chinook. We've got to do some major, major work on habitat, we've got to rebuild genetic diversity, and we've got to stop non-selective fishing once and for all."

As the bycatch issue in salmon net fisheries grew more controversial during the 1990s, DFO was finding it harder and harder to secure the cooperation of some increasingly nervous sections of the fleet. In 1994, for instance, during seine fisheries for 600,000 chum salmon in Juan de Fuca Strait off Nitinat, on Vancouver Island's west coast, fishing companies supplied DFO with catch information that indicated their boats had encountered not a single coho, steelhead or chinook salmon as bycatch. For at least the preceding ten years, however, the bycatch for coho alone during the Nitinat chum seine fisheries had been reported to average about 40,000 fish per year. The discrepancies prompted the Ditidaht First Nation to plead with then Fisheries Minister Brian Tobin for a mandatory observer program during the seine openings.

A similar example of non-cooperation by commercial fishers also comes from the Nitinat chum fishery. In 1991, DFO's routine assessments of the bycatch of non-target salmon species during three chum gillnet openings were compared with reports from provincial government staff who were monitoring chum deliveries to fish-packing vessels on the fishing grounds. While DFO had been provided with information suggesting a bycatch during the fisheries of only 1,407 coho and 10 steelhead, data assembled by provincial biologist Rob Bison indicated a bycatch of 3,790 coho and 236 steelhead. By 1995, there was still no bycatch monitoring program in force for this fishery.

In a September 29, 1995, letter to Louis Tousignant, regional director of DFO's Pacific region, Fred Fortier, chairman of the Shuswap Nation Fisheries Commission, protested the Juan de Fuca chum fishing plans, pointing out that not only endangered U.S. coho but also steelhead and coho bound for streams in the Fraser and Thompson watersheds would be unduly impacted. At stake were distinct coho populations that had supported the traditional fisheries of the Bonaparte, Kamloops, Adams Lake, Neskonlith, Little Shuswap, Skeetchestn and Spallumcheen bands. The Skeetchestn band had voluntarily closed its steelhead fisheries for ten years, and in their efforts, the communities were resorting to "gene banking" of local coho stocks with the help of the World Heritage Trust—an effort that was "not a solution," Fortier noted, but a "last resort."

Fortier strongly urged Tousignant to restrain the non-selective fishing technologies deployed on the coast: "We expect that the surviving commercial fleet in the future will see the wisdom in sustainable, selective fisheries, and can assure them that we would support their unimpeded fishery if it too were designed to be selective of the target stock, and respectful of aboriginal rights."

There was no small irony in this. Coyote himself, after he had broken the dam and allowed salmon to return to their home rivers and streams and creeks, instructed these same Shuswap people in the use of selective fisheries.

But that was in the distant past, somewhere around the beginning of time. Coyote was gone, and Mechanism was in his place. By the time Tousignant read Fortier's letter, more than two hundred gillnetters were heading out to Juan de Fuca Strait to set their nets, during the first week in October, in the waters off Nitinat.

The seiners went next.

CHAPTER V

⚓

Coyote's Fall

Rational salmon management is not just a search for technologies:
It is a search for values.

—D. L. Scarnecchia,
"Salmon Management and the Search for Values"

Nature is only to be commanded by obeying her.

—Francis Bacon

AS THE CLOSE OF THE TWENTIETH CENTURY LOOMS, THE MOST PRO-
found threat the commercial fishing industry perceives to its security is
the spectre of aboriginal fishing rights.

Nothing has sent more shudders through the industrial fisheries
edifice than the prospect of an emerging Indian challenge to the
industry's 90 per cent share of the coastwide total allowable catch of
salmon. While the 1994 Task Force on Fish Processing noted an
"evolving jurisdiction" of First Nations in fisheries under Section 35 of
the Constitution Act that was "not yet fully defined," a 1993 B.C.
government survey of industry participants' views about the state of
the industry concluded that the prospect of increased tribal allocations
was regarded as "the single greatest threat to the B.C. commercial
fishery."

The 1993 survey revealed deeply felt insecurity about "aboriginal
claims," noting that "many question the need for further government
assistance to aboriginals." Industry participants feared they were being
"asked to bear the costs of settling difficult aboriginal claims," and

they feared "losing access to a sizeable portion of the resource they depend on without compensation or any say in the matter." Pilot sales programs, initiated in some in-river tribal fisheries formerly confined to food fishing, were perceived as a threat to B.C.'s reputation for high-quality salmon. Robert Heywood and Timothy Taylor, the authors of the 1993 study, reported that such initiatives in the tribal fisheries were regarded as presenting new competition: "Aboriginal communities may well wish to establish their own processing capacities to take advantage of their newly guaranteed supply of fish," resulting in "insecurity" and a hampering of the industry's ability to compete on international markets. Further, aboriginal rights were seen as an ominous development that could "dangerously supersede DFO's right or will to manage the fishery in the best possible way."

It may be that in 1993 the spectre of aboriginal fishing rights was producing particularly acute concerns among rank-and-file commercial fishers (there were ear-splitting "missing fish" alarums in 1992, with "massive native poaching" widely identified as the culprit), but the prospect of a revival of tribal fisheries had been causing extreme anxieties throughout the industry for several years.

Historically, the fishing industry and the federal government have approached the presence of aboriginal fisheries with the intention of restraining them, eliminating them altogether or integrating aboriginal people, as individual wage earners and harvesters, into the industrial fisheries complex that emerged on the coast in the mid-1800s. Any resistance to that policy, and any minor revival of the tribal fisheries complex, have been met with fierce opposition. By 1993, highly effective media campaigns and industry lobbying efforts against native fisheries were being coordinated by the B.C. Fisheries Survival Coalition, but the coalition itself was only the latest version in a series of industry initiatives that had gone under several names, most notably the Pacific Fishermen's Alliance, which later became the Pacific Fishermen's Defence Alliance. Among several fishing organizations, the Alliance enjoyed its own intervener status at the Supreme Court of Canada where it fought—and lost—the 1990 confirmation of aboriginal fishing rights as protected rights under Section 35 of Canada's constitution, in the Sparrow decision.

From 1986 to 1990, the Alliance and the Fisheries Council of B.C. had intervened in the courts against aboriginal rights on several occasions. They tried (and failed) to win a court injunction against any tentative agreement covering fisheries in the Nisga'a comprehensive claims negotiations. They intervened (and won) in a case against Indian band bylaws on the Skeena River in 1986 that would have allowed a limited, commercial in-river fishery.

The complexity of the conflict over aboriginal fishing rights cannot be easily boiled down to two opposing arguments. But if there is an aboriginal side to the argument, it was expressed well in an analysis written by Calvin Helin, a Tsimshian lawyer and member of a prominent north coast fishing family. In 1993, he wrote: "In light of the devastating social change that has taken place in First Nations over the last two centuries, it is ... difficult to feel very sorry for powerful corporate conglomerates that have taken jobs out of native communities by centralizing the processing industry in urban centres. It is difficult to sympathize with the commercial interests that can afford to monopolize and concentrate their power through access to large capital pools that are unavailable to aboriginal people." Helin's advice to those who oppose tribal fisheries is plain: "In attempting to come to grips with Sparrow, many of the detractors of aboriginal fishing interests might well be advised to keep the comments of Barnett J. in mind. In R. v. Bones, he said: 'Some other people say that the courts want to give aboriginal people total control over fish, wildlife, and other resources across Canada, and that if this happens our resources will soon be destroyed. Persons who make such statements are mischievous scaremongers who have not read or understood what the courts are saying.'"

The opposing side is represented mainly by B.C.'s commercial fishing industry. From the most conservative company president to the most progressive union local, the argument has been for "industrial solutions" and affirmative action to satisfy aboriginal demands: Let them fish as individuals, within the existing commercial fishery structure, taking their risks in the unstable harvesting sector along with everybody else; if they persist in their tribal fisheries, let them confine their activities to food fishing. There is nothing unreasonable about these arguments, given the values and the economic interests that produced them. Besides, depending on whose statistics you believe, by the 1990s it may have been that as many as 30 per cent of the fishers, crew members and seasonal cannery workers involved in the commercial fishing industry were of aboriginal descent. And on years of low sockeye returns to the Fraser River, when other coastal rivers provided the bulk of the coastwide sockeye supply, the Fraser's in-river tribal catch was coming close to half as much as the saltwater, commercial Fraser salmon catch. In 1988, Fisheries Council president Mike Hunter articulated quite succinctly the dominant industry view of aboriginal rights: "Obnoxious, invalid and un-Canadian." In 1990, Fisheries Alliance lawyer Chris Harvey warned of a "back to the jungle free for all" should natives regain substantive control over their own fisheries.

When native communities pursue their own course, often defying fisheries regulations along the way and often finding support from the courts or public opinion or government policy in their aims, it sometimes becomes more difficult to describe the fishing industry's response as reasonable. At times—with due regard to the notable exceptions—the industry's response has bordered on hysteria. It has been as though Mechanism is suffering nightmares that Coyote has returned to break the dam and is slowly floating downriver, headed for the coast.

It would not be true to say that the federal government, the fishing companies, the fishermen's union and all the rest are recklessly destroying the fishery. But it is a fact of modern public policy that there is desperately scant evidence that the prevailing industrial fisheries management system can be made to work properly. The large-scale industrial management system governing salmon fisheries is not just an engine in need of an overhaul. Throughout the late 1900s, the Pacific fisheries and the management systems that ostensibly governed them were constantly being recalled or hauled into the repair shop for overhauls. Yet nothing worked.

Similarly, there is not some idyllic aboriginal paradise to which we might all simply return. Any suggestion that fisheries management is a responsibility that should be simply returned to "the Indians" along with the catch is as reckless as it is naive. But there remains one significant thing that distinguishes the old aboriginal salmon management systems from the industrial fisheries that took their place: they worked. And they worked for a long, long time. For these reasons alone, aboriginal fisheries warrant a fairly close study in any search for new fisheries management models.

There was Coyote, but there was also Raven, and there was Salmon Boy and there was X:als. These mythologies did not produce fisheries that were quaint little affairs undertaken by the timid children of the forest, as some environmentalists appear to imagine. Neither is it true that these fisheries were the crude handiwork of primitive savages whose only boast might be that they didn't destroy the resource because they didn't have the wherewithal to do it. The reality is that there were at least tens of thousands of people involved in directed fisheries upon a variety of fish populations throughout the British Columbia coast, and well into the interior, for countless centuries before disease, dislocation and federal fisheries regulations came on the scene. The oral tradition, early explorers' and traders' accounts and the archaeological record all show that these fisheries were sophisticated and substantial undertakings, involving volumes of salmon perhaps not far short of the average annual harvests the twentieth century's industrial fisheries have produced.

Archaeological evidence suggests that the advent of the Fraser basin's tribal fisheries occurred as far back in the mists of time as fishing cultures go anywhere on the planet.

Conventionally, the Earth's first "fishing cultures" are held to have been a Mesolithic people who were gathered in large villages around the Baltic Sea, from about 8,000 years ago to 4,000 years ago. These people have come to be known as the "Maglemosians," a word derived from *Maglemose*, a Danish term meaning "big bog," which refers to the kind of environment in which their settlements and shell middens have been found. Throughout the world at about the same time, sedentary communities of people—localized communities, more like agricultural societies than classic "hunter-gatherer" peoples—were establishing themselves adjacent to marine resources, and harvesting the resources of the land as well, at places such as the mouth of the Nile River on the African continent and at Baja, California. The Jomon people of early Japan are among the most notable of these cultures, and there has been some speculation that trans-Pacific cultural diffusion during these early times might have provided some tenuous links between Jomon peoples, Kamchadals, Sakhalin Islanders, Aleuts, "Eskimos," Tlingits and other peoples on the North American side of the Pacific Ocean.

Roughly at the same time the Maglemosians were trolling with barbed fish hooks and skiffing about the Baltic shoreline in rough-hewn watercraft, a flourishing community of fishing people was emerging on a bluff adjacent to a vast estuarine marsh in what is now North Delta, on the Fraser River. At the "Glenrose" site, which now lies in the shadow of the Alex Fraser Bridge, archaeologists have unearthed fish and animal bones 6,000 to 7,000 years old in heaps of mussel and clam shells. Back then, Glenrose was a seaside village at the mouth of the river. But by the early 1900s, the Fraser River had formed wetlands far to the west, and by the late 1900s, after a century of draining and dyking, the mouth of the river was several miles downstream. Elsewhere, on the rivers and along the coastline of what was later called British Columbia, other fishing cultures were establishing themselves, placing B.C. aboriginal societies among the first peoples on the planet to adopt maritime and semi-sedentary ways of life. Almost immediately after the glaciers receded, at about the same time salmon were re-colonizing the land, human communities arrived and began the long process of coevolving with salmon populations.

Throughout the Fraser basin, the fishing technology that may have produced the greatest volume of fish was the fish weir, a fencelike structure built out into a river or completely across a river, with staves that were opened at the end of a fishing day. Fishing technologies employed in the lower reaches of the Fraser River, among the Halkomelem-speaking peoples, are well described by anthropologist Gordon Mohs, whose inventory of catch technology cites nets, traps, weirs, hooks and harpoons. Net types were primarily dipnets and bag-nets. Dipnets were employed most effectively in the canyon, from ledges, outcrops and platforms built out over the river. Bag-nets were used in trawling the lower sections of the river. Mainstem Fraser fishing among the Nlaka'pamux, the Lillooet and the Secwepemc was carried out with a number of devices, but primarily by the use of dipnets, not unlike the Sto:lo dipnet described by Mohs. This same technology was used in the fisheries on the Chilcotin River. Large schools of salmon congregate in the back eddies of the Chilcotin as they fight the current on their way upstream. At these sites, dipnetting was undertaken, sometimes from platforms or log "stagings" constructed out over the river. Some dipnets were made of mesh fixed directly to a hoop at the end of a long pole; bag-nets were fixed to retractable horn rings, allowing the fisherman to close the net like a purse when a salmon was caught.

The importance of salmon in the diet and the economies of upriver aboriginal communities has tended to be overlooked by historians and fisheries management officials from the earliest days of colonization. In a July 30, 1878, letter to E. A. Meredith, minister of the interior,

Reserve Commissioner Gilbert Malcolm Sproat tried to correct this misconception, observing, "An impression exists that there are two classes of Indians in this Province, namely those who live on fish and those who live on flesh. This is only true in the sense that the Coast Indians eat many kinds of fish, and the Interior Indians eat only salmon and trout, while both Coast and Interior Indians vary their diet occasionally with grouse, ducks and venison. The whole 30 or 40,000 Indians in British Columbia are a fish eating people, and to all of them, but especially perhaps to the Interior Indians, the salmon is the principle article of fish diet ... 'The salmon first and God next,' an old Indian said to me last year."

The false assumption Sproat argued against persists, however. It is a widely held idea, even though there has never been any evidence to support it, that coastal aboriginal peoples relied more heavily on salmon than did interior, upriver peoples. As the recent research of University of British Columbia archaeologist Brian Chisholm has shown, the opposite may well be true, given the extremely high productivity of fishing sites throughout the Fraser basin. By analyzing skeletal remains, Chisholm has studied the marine component of the protein in aboriginal people's diets (salmon protein is discernible in these analyses from other types of protein, including freshwater aquatic protein), and his findings suggest that salmon comprised at least 50 per cent of the protein component of First Nations' diets well into the upriver reaches of the Fraser basin.

In the Fraser Canyon, one important archaeological site, known as the Milliken site, offers persuasive evidence of salmon fishing nine millennia into the past. Charred pits of the wild cherry, which ripens in August and September when sockeye are running the thickest through the Fraser Canyon, have been found in great quantities at the Milliken site, and archaeologist Knut Fladmark notes that the evidence at Milliken strongly suggests that communities were organizing their seasonal movements around the salmon cycle 9,000 years ago. From the research of Gordon Mohs, there is no question that efficient techniques for harvesting and processing quantities of salmon were in place in the Fraser Canyon area 5,000 years ago. The earliest certain evidence of salmon harvesting in the Fraser system lies about two hundred miles upstream from saltwater, in the Nlaka'pamux territory, near the confluence of the Nicola and Thompson rivers. People were fishing there at least 7,500 years ago. At several other locations throughout the Fraser plateau, archaeological excavations have produced evidence of specialized net-fishing technology at least 4,500 years ago.

A study for the Westwater Research Centre by Michael Kew and Julian Griggs sheds some light on the dispersion of human communities throughout the Fraser basin from earliest times. The study found that by the late 1980s, almost every remaining village of the ninety-plus Indian bands in the Fraser basin, each of which belongs to one of several formerly larger cultural groups, was located at or near a fishing site within a particular tribal territory.

At the most northerly extent of the Fraser basin there are Athapaskan-speaking Carrier peoples. While the Fraser River salmon runs available to Carrier peoples for harvest are relatively few, they have been among the Fraser's largest, such as the Stuart Lake stocks. Present-day bands include Red Bluff, Nazko, Kluskus, Ulkatcho, Fort George, Stoney Creek, Nadleh Whuten, Stellaquo, Chesletta, Broman Lake, Burns Lake, Necoslie, Tl'azt'en and Takla Lake. Moving south and southwest, the Ts'ilhqot'in (Chilcotin) people are also considered Athapaskan, and they direct their fisheries primarily in the Chilcotin River. The present-day Chilcotin communities range from Alexandria in the east to Nemiah in the southwest. Moving south and east, the Secwepemc (Shuswap), classified along with the Nlaka'pamux and the Lillooet as "Interior Salish," occupy perhaps the largest land mass of any other First Nation west of the Rockies. The Secwepemc traditionally enjoyed direct access to the mainstem of the Fraser River, along with the Thompson River and its tributaries. There are seventeen Indian bands within the Fraser basin that identify as Secwepemc, from Spallumcheen in the southeast to Soda Creek in the northwest. The Lillooet people, including the Stl'atl'imx, the Lakes Lillooet and the Mount Currie people, have traditionally occupied key fishing areas in the central watershed, including a section of the mainstem Fraser River, the Bridge River, the Lillooet River and the Birkenhead River. Lillooet fisheries contributed substantially to far-flung networks of trade in dried salmon. Their present-day communities include eleven Indian bands. The Nlaka'pamux (Thompson) occupy a crucial portion of the mainstem Fraser River, as well as the Nicola and other smaller rivers. Their access to the Fraser Canyon produced some of the most intensive and productive fisheries in the entire Fraser River watershed. Present-day communities include sixteen bands from the

Nicola Lake area in the northeast to the vicinity of Spuzzum in the southwest.

At Spuzzum, the Nlaka'pamux southern frontier marks not only its "border" with the Upper Sto:lo people but also the division anthropologists and linguists have drawn between the "Interior Salish" and the "Coast Salish" areas, as well as the boundary between the larger "Fraser River Plateau" and the "Northwest Coast" culture area. From a point on the mainstem Fraser in the vicinity of Spuzzum to the mouth of the river and the sea coast, dialects of Halkomelem are spoken. There are twenty-eight Indian bands in this area, which was once one of the most populous regions of the Pacific slopes of North America. It may be that these Lower Fraser peoples consumed more salmon per capita than any other tribal community in the Fraser basin; they were certainly located strategically to harvest portions of all the salmon species and all the species' runs bound for their natal streams above the canyon, and they were known to rely heavily on salmon bound for smaller rivers and streams in the Fraser basin below the Fraser canyon. The overwhelming majority of the Lower Fraser people, in more than twenty separate bands, identify as Sto:lo, but they include also Coquitlam, Katzie, Musqueam and Tsawwassen.

Without doubt, there were numerous coastal aboriginal communities that fished Fraser-bound stocks at least casually, particularly chinook and coho, which were taken easily with aboriginal trolling gear. But there were also numerous communities with linguistic, cultural and family ties to Lower Fraser tribes that were involved in directed fisheries upon Fraser River stocks, either by seasonal visits to the Fraser River or by the use of specific fishing technologies in saltwater "approach" fisheries. The communities with traditions of directed fisheries upon Fraser-bound salmon include Vancouver Island and Gulf Islands communities whose dialects fall within the Halkomelem language. They include the Cowichan, the Nanaimo, the Nanoose, the Penelakut, the Lyacksen and the Chemainus peoples. These communities sent flotillas of canoes to fish the Fraser estuary, and some travelled as far upstream as the Fraser Canyon to participate in the intensive fisheries there. Similarly, the Squamish and the Burrard people appear to have enjoyed long-standing "licence" to Fraser River stocks as well. The Straits Salish peoples, meanwhile, also conducted directed fisheries upon Fraser River stocks by utilizing a form of the saltwater, live-capture reefnet technologies that were unique on North America's West Coast, usually deployed around jutting headlands and points, and directed upon thickly schooling runs of salmon, usually from the same species and stock grouping. Straits peoples include the bands of the Saanich tribe, including Semiahmoo and Songhees, as well as the Samish and the Lummi peoples in what is now Washington State.

The preceding roll call helps to illustrate some key characteristics that distinguish pre-contact aboriginal fisheries from contemporary salmon fisheries. As often as not, pre-contact fisheries tended to be stock-specific or run-specific, taking place mainly within the Fraser River, as far upstream as the terminal areas, on primary tributaries and on secondary and tertiary rivers and streams. Fishing effort was widely dispersed along the salmon migratory path, and fisheries were governed by locally based systems of access rights. Mixed-stock fisheries at the beginning of the fishing sequence may have produced high volumes of fish, but they represented very low harvest rates; fisheries became more intensive and more productive further along the salmon migratory path, becoming increasingly stock-selective and species-selective the further up the Fraser and its tributaries they were, allowing substantial catches at higher harvest rates the more terminal the fishery's location.

The industrial, ocean-based fisheries that developed in place of the old patterns quickly ended up looking almost exactly the same as the aboriginal system—only turned upside down.

Critics of such comparisons say it is false to hold that scores of distinct aboriginal fisheries, traversing hundreds of kilometres, occurring across several language barriers and mountain ranges, can be properly described as a managed system. This is a legitimate criticism.

But there is voluminous anecdotal evidence of communities throughout river systems jealously guarding their fishing stations, ever watchful of the fishing habits of communities downriver, and the customary laws of tribal communities may have played an important role in controlling fishing effort and regulating the catch throughout much of the Fraser basin. Similar forms of customary law were undoubtedly involved in the regulation of access rights to the Fraser mainstem by Island Halkomelem and other communities. Such complex characteristics may not amount to a purposefully managed system at all. But the point is that these salmon populations, and the human communities that interacted with them, seem to have flourished perfectly well for thousands of years before the Department of Fisheries and Oceans was established, and the real point at stake here is that these fisheries worked. The evidence is in the presence of scores of ancient fishing communities dispersed throughout the Fraser basin, and the presence of a spawning salmon population in virtually every accessible creek and stream of the Fraser basin well into the 1800s.

Another criticism of the usefulness of considering aboriginal fishing patterns has been that river-caught salmon have less commercial appeal than "bright" ocean-caught fish, and that they deteriorate quickly once they enter fresh water. This is indeed a problem, but it is only true of some stocks, and many salmon caught well into the river would probably be indistinguishable from ocean salmon to all but the most astute consumer. Generally, it is only those fish caught particularly close to their natal streams that deserve their "gumboot" reputation, and most of these criticisms ignore the potential value that could be added by the production of traditional commodities that formerly fed large communities of people throughout the long winters: smoked, smoke-dried, sun-dried and wind-dried salmon.

A more relevant question about the usefulness of considering aboriginal fishing patterns concerns the actual numbers of fish such a management model might produce in a modern context.

Just how many Fraser salmon were regularly caught during the aboriginal period is a difficult question to answer, although several scientists and historians have tried. The veteran fisheries scientist William Ricker has suggested that the pre-contact aboriginal catch of Fraser River sockeye salmon could have amounted to "several million" fish annually, and it is important to note that all species of salmon were harvested in the Fraser's tribal fisheries. Chinook was preferable to sockeye among many tribal communities, and chum salmon was the prime smoking fish to be stored through the winter. In 1973, Gordon Hewes attempted to calculate what that salmon catch might have been by what he describes as rather "arbitrary" speculation about the caloric requirement of First Nations peoples on the Pacific slopes of North America—2,000 calories per day per person. Hewes then goes on to suggest this requirement would amount to roughly two pounds of food per day, of which half would likely be salmon, producing an average estimated per-capita requirement of 365 pounds of fish per year. Since a sockeye weighs about five pounds on average, such a requirement would equal seventy-three sockeye per year, per person, or the equivalent in smaller fish such as pink salmon or larger fish such as chinook, which often weigh in excess of forty pounds. Hewes's estimates of pre-contact Fraser River First Nations salmon consumption: Fraser Delta, 1,000 pounds per capita; Ts'ilhqot'in, 600 pounds per capita; Lillooet, 600 pounds per capita; Thompson (Nlaka'pamux), 900 pounds per capita; Shuswap (Secwepemc), 500 pounds per capita; Carrier, 600 pounds per capita.

Among the Lillooet, Brian Hayden summarizes the research of Steven Romanoff, Dorothy Kennedy and Randy Bouchard this way: "It appears that families with proper access could obtain at least 300-400 spring [chinook] salmon and 200-300 sockeye per year (about 9,000 pounds total, or 1,800 pounds per person for a family of five, or 5 pounds per person per day)." This amounts to an equivalent of about 365 sockeye per capita annually, given that sockeye are roughly five pounds in weight on average. Not all of these fish would have necessarily been consumed directly by Lillooet people; some might have been exchanged in trade. The Lillooet estimates are significant when one considers that on a per-capita basis, if the 4,000 or so Lillooet were people catching those amounts of fish in 1995, they would be due some 1.4 million salmon annually, a figure well in excess of the entire contemporary annual allocation of Fraser salmon to all Fraser River First Nations and all saltwater Indian bands, combined.

Research clearly shows that even those "remote" interior First Nations not generally associated with intensive salmon harvesting traditions, such as the Ts'ilhqot'in, relied heavily on salmon throughout the centuries, perhaps every bit as much as coastal peoples. In his 1984 ethnographic study of the Ts'ilhqot'in people, Robert Tyhurst found that "Fish, of whatever type, was undoubtedly the single most important source of food for all Chilcotin until well into the latter half of the 19th century." The Ts'ilhqot'in named three months of the year—July, August and September—after chinook, sockeye and pink salmon. The same reliance on salmon can be found among all the Carrier peoples. The western Carrier people, the Wet'suwet'en, directly utilized the rich salmon resources of the Skeena and Bulkley rivers. The Fraser basin Carrier obtained fish in trade from their western neighbours to some degree, but they engaged in substantial fisheries of their own on the mainstem Fraser River, the Stuart, the Nechako and the many smaller Fraser River tributaries in their territories, and their fishing technology was as elaborate and varied as any to be found throughout the Fraser basin: fish weirs, conical fish traps, wooden fish rakes, three-pronged spears and nets of willow and alder bark.

In the end, the total production that might be attributed to the Fraser's aboriginal fisheries depends on how many people were involved in them.

The anthropologist Wilson Duff was the first to attempt calculations of West Coast aboriginal populations, and his estimates suggest only about 24,500 people among all Fraser River tribal groups by 1800, taking into account the population declines that had by then already devastated many communities. (Epidemic diseases, inadvertently introduced by Europeans, had swept across the continent well in advance of white explorers.) Post-colonial epidemics contributed to further declines among the Fraser tribes: 22,200 people by 1835 and the nadir of 8,046 by 1926. From 1929 the population began to recover, in many cases quite dramatically, and by the late 1980s at least 27,000 "status" Indians were registered among ninety-plus bands in the Fraser basin.

While Duff identified the Sto:lo area and the neighbouring Nlaka'pamux area as two of the most populous regions of the northwest corner of the North American continent prior to contact with Europeans, anthropologist Gordon Mohs argues that the Upper Sto:lo population, before the outbreak of disease epidemics, was "upwards of 8,750-28,000," as opposed to Duff's estimates of 2,500 to 3,500. The Carrier peoples are generally held to have enjoyed the highest population density of any "Subarctic Athapaskan" group, at about 8,500 people, according to Margaret Tobey's reports to the Smithsonian Institution. Among the Secwepemc, according to anthropologist Marianne Boelscher-Ignace, a population of 9,000 was possible.

The explorer Simon Fraser, who kept a journal during his travels in 1808 down the river that would be named after him, provides a glimpse of what the aboriginal population might have been at that time, although even at that date First Nations populations may very well have been severely diminished because of at least one smallpox epidemic. (It may also be that Fraser's visit brought disease again, further diminishing the population.) Among the Nlaka'pamux, Fraser recorded a visit to the village of Cumsheen, at what is now Lytton. After leaving a village of four hundred people on the west bank of the river, he arrived at Cumsheen, where the chief "took me by the arms and conducted me in a moment up the hill to the camp where his people were sitting in rows to the number of twelve hundred; and I had to shake hands with all of them." Fraser and his men were then treated to a feast of salmon, berries, oil, roots and dogs. Fifty years later, during the gold rush of 1858, white miners reported the Nlaka'pamux population along the Fraser River between Cumsheen and Spuzzum to be about 2,000, suggesting a total Nlaka'pamux population by that time of about 5,000. At the turn of the twentieth century, the ethnographer James Teit noted that "the old Indians compare the number of people formerly living in the vicinity of Lytton to 'ants about an anthill.'"

Whatever population levels may have been prior to introduced diseases, there are two statements most anthropologists and archaeologists agree upon: Duff's assertion that "No other region of Canada was so heavily occupied; in fact about 40 per cent of all the native people of the country lived within the present boundaries of British Columbia," and geographer Cole Harris's conclusion that "The Fraser was a huge source of food, and in the canyon, where fishing sites were abundant, and excellent, it probably supported as concentrated and dense a non-agricultural population as anywhere in the world."

In the end, we are left with guesses about what the total salmon catch might have been. However, if the aboriginal population of the Fraser basin in the early 1700s was somewhere around the median between Duff's conservative estimates and the estimates of Mohs, Boelscher-Ignace and Tobey (say 60,000 people), and if their fisheries amounted to no more than the median estimated consumption rate of 800 pounds per capita, per annum, or the equivalent in salmon of 160 sockeye per person, per year, the combined Fraser River tribal catch would have amounted to about 6 million five-pound salmon per year—equivalent to about 6 million sockeye. If Romanoff, Kennedy and Bouchard are right about the Lillooet fisheries, and harvest levels are applied proportionately throughout the basin, the annual Fraser River pre-contact tribal catch could have been as much as 12 million sockeye equivalents. The most recent estimates of this kind come from Westwater's Thomas Northcote, in his efforts to establish the requirements for sustainability in fisheries, fish habitat and human populations within the Fraser basin. Based on an aboriginal population estimate of 50,000 people, Northcote reckons on an annual harvest of 5.3 million five-pound salmon. By way of comparison, the average total

abundance—the all-sector catch plus spawning escapement—of all
Fraser River sockeye runs between 1894 and 1986 ranged only from
3.31 million to 11.17 million fish.

These pre-contact catch estimates do not begin to account for the
more casual harvests of Fraser-bound salmon by communities adjacent
to Fraser salmon migratory routes, or the catch by communities such
as the Island Halkomelem tribes with traditions of travelling to the
Fraser to conduct directed fisheries on the river's lower mainstem, or
the catch taken in the elaborate, directed reefnet fisheries maintained
by Straits Salish peoples.

If there were a final chapter on the tribal fisheries of the Fraser
basin, it would have to open with Clause 1 of the British Columbia
Fishery Regulations, Dominion Order in Council No. 590, 3 March
1894. With this regulation, fish weirs and other aboriginal technologies were outlawed as part of the regulatory scheme established
ostensibly to govern the coastal canning monopoly. The 1894 regulations also reasserted seldom-enforced rules enacted a few years earlier
that prohibited aboriginal people from selling their fish unless they
travelled to the coast, obtained a permit from a cannery and sold the
fish to the cannery. The 1894 rules stated that upriver natives could
fish only for their food, and only with the permission of the inspector
of fisheries, but not with traps or "pens." The new rules were hotly
protested by native peoples, and the B.C. government formally joined
their protests in a July 1894 order-in-council, warning Ottawa that it
was provoking a violent response from Indians.

The provincial order-in-council states that apart from employment
obtained by some Indians in the coastal salmon canneries, "few have
any means of obtaining ready money, the only source being the sale of
such game and such fish as those resident near populous centres are
able to find a market for." But by 1894, thousands of native people
were already travelling to the sea coast every year to work for cannery
wages or to fish directly for the canneries. They travelled to the mouth
of the Fraser every season, from as far afield as the upper reaches of the
Fraser basin and as far up coast as the Queen Charlotte Islands.
Throughout the province, meanwhile, fisheries officials were dismantling traditional fishing technologies on the Fraser, the Nass, the
Cowichan, the Skeena and the Babine. Their campaign has been
meticulously recorded by historian Reuben Ware in his 1983 document
Five Issues: Five Battlegrounds.

The coastal fishing industry had opposed aboriginal technologies years before weirs were finally outlawed, arguing that weirs destroyed salmon fry and prevented adult salmon from spawning. The controversy prompted the following discussion at a reserve-commission hearing at Okanagan Lake on September 24, 1877. Secwepemc interpreter Antoine Gregoire, described by Indian land commissioner Gilbert Malcolm Sproat as a "competent and trustworthy witness," was questioned by Alexander Caulfeild Anderson, B.C. fisheries inspector. With regard to the destruction of salmon fry, Gregoire says "positively that it does not take place ... They know, and say, that if the young fish are destroyed, the shoals returning from the sea will be proportionately diminished." On weirs: "The barriers they construct in rivers are only to retard the passage of the fish, to enable the Indians to obtain their necessary winter supply, and ... these temporary obstructions are thrown open, as necessary, to give passage to the ascending fish." As for allegations that Indians destroyed the spawn in the beds, Gregoire was asked whether he had ever heard of such a practice. His response is reported this way: "Never. Thinks that the whole statement is imaginary."

Gregoire adds that the chiefs were so careful of the salmon that they would not allow their people to use poles from their canoes to traverse rivers containing spawning beds; only paddles were permitted, and the customary laws were such that children were prohibited from building toy weirs to trap juvenile salmon in the spring.

Some weirs were already being demolished a year before fisheries regulations formally prohibited them. In 1893, fish weirs on the Nicola River were destroyed. In other places, weirs remained in use years after the 1894 regulations banned them. At Bridge River Falls, weir fisheries had been extremely important for the local Stl'atl'imx people. The weirs there were destroyed for the last time in 1912, ending a protracted dispute between tribal and federal authorities.

David Salmond Mitchell, one of the first federal fisheries officers in the Fraser basin and a key figure in later government initiatives aimed at restoring damaged salmon runs, was a firm defender of aboriginal weir fisheries. In his 1925 memoir, Mitchell reported: "Their barricades across the streams were opened at the ends as soon as they had as many fish as the women could prepare for drying that day. They were opened, as the Indians said, to let salmon go upstream to 'mammok papoose' (jargon for reproduce) ... I have gone to their barricades at all times, unexpectedly, and in the middle of the night, and found that it was so."

Sensationalistic headlines—common today in industry-inspired agitation against the resurgence of tribal fisheries—were the rule at the turn of the century as well, with one headline from the *Province* newspaper of November 19, 1904, reading, "Indians Wiping out Sockeyes." Subsequent investigations invariably proved such allegations to be groundless.

But nothing was so devastating to the Fraser basin's tribal fisheries as the events that were to come—events that began in the years 1911 to 1915, a time of massive commercial harvests of Fraser-bound salmon and of the westward expansion of Canada's railway systems.

Hell's Gate, an extremely narrow gorge in the Fraser Canyon about 130 kilometres from the river mouth, had always been a particularly difficult and potentially deadly obstacle in the path of salmon migrating home to their upriver spawning grounds. By the 1880s, crews building the Canadian Pacific Railway had already blasted thousands of tons of rock into the gorge, making salmon passage even more difficult. But between 1911 and 1912, Canadian National Railway crews fairly exploded the side of a mountain into the canyon. And in 1913, the commercial catch of Fraser River sockeye was an estimated 32 million fish, a commercial catch greater than any before or since—in fact twice as many Fraser sockeye as the commercial fishery has produced in any year since then. When the pathetic remnants of the 1913 runs arrived at Hell's Gate, they massed below the canyon, unable to pass. Only a few of the most hardy fish survived their attempts to traverse the gorge. Then, in February 1914, an estimated 100,000 cubic meters of granite collapsed into the river from a huge rock bluff construction crews had punched a tunnel through. One of the largest rivers in the world, draining an area larger than England, had come meters away from being entirely dammed by accident. Immediately, in the gathering winter, deep within the canyon, provincial and federal engineers worked night and day, blasting chutes and building wooden flumes, shouting their instructions above the roar of the river. Aboriginal people set to work dipnetting salmon from below the carnage, one at a time, rushing them to rickety, hastily constructed timber viaducts that carried the fish to tentative safety above the falls. The operation was like some desperate, wartime field hospital. With increasing degrees of effort and sophistication, the operation lasted for several years.

It is impossible to describe the depth and breadth of the hardship these events visited upon the tribal communities of the Fraser basin. Glimpses of what it must have been like survive in stories native elders tell, and in various reports and archival records. In small, remote native communities there was hardship approaching famine. Large mainstem communities were places of tremendous grief, shock and disbelief. To avoid starvation, many people were forced to leave their communities, becoming refugees in their own lands. While fisheries continued on the coast, the few salmon that did survive their homeward migrations through the canyon were all required for spawning, forcing fisheries officers to travel throughout the B.C. interior to enforce rigid, unprecedented fishing closures, sometimes prohibiting fisheries entirely.

Brief snapshots of the hardship of these times emerge from the testimony heard by the Royal Commission on Indian Affairs (the McKenna-McBride Commission) during the years 1913 to 1916. Fountain Chief Tommy Adolph spoke of his community waiting all summer for salmon to arrive at their fishing stations. Word of a few salmon reached the village, "and when the Indians went down to catch them, we were stopped." Mount Currie Chief James Stager told of steady, rapid declines in salmon abundance, and the depopulation of his village as a result: "Three quarters of my people are away for their living because they cannot get sufficient Salmon... Last winter, four people had to live on seventy Salmon, so all my people are very sorry about this bad state of affairs. Some of my people here have five in a family and sometimes six. Now sometimes we don't get one Salmon." Fort George Sub-Chief Joseph testified that his community had been driven away from the Fraser River; they couldn't find fish in the Nechako River and they weren't allowed to fish in the Salmon River: "The Indians had no fish and no places to fish ... if we go there now, we will be chased away." Stellaquo Chief Isodore, Necoslie Chief Jimmy and Necoslie spokesman Sam Prince all protested the prohibition against their weir fisheries and reported that gillnets provided by government officials were coming up empty. As a result, the people "sometimes starved." At Boston Bar, the Royal Commission heard this testimony, from a local chief: "The reason of this slide [at Hell's Gate] is caused by white men ... I don't want to be stopped from fishing salmon in the river. God made those for our use, and it is from salmon that I make my living. Therefore, I wish everything to be free."

As historian Dianne Newell has documented, the tribal fisheries restrictions that followed Hell's Gate evolved into measures designed to eliminate the Fraser basin's tribal fisheries entirely, despite the fact that those fisheries continued only as remnants of their former significance. B.C.'s chief inspector of fisheries, James Motherwell, was one of those who suggested a complete and final prohibition on the tribal fisheries, and his position was consistent with the views of the B.C. Salmon Canners Association, which by the late 1920s was already arguing that "the [food] catch of the Indians should be curtailed if not eliminated entirely." These moves came at a time when the entire aboriginal population throughout the Fraser basin had been reduced to fewer than 10,000 people.

During the 1930s, the canners sponsored attempts to wean aboriginal communities off salmon completely by providing communities with canned pilchards, and in 1931, the market for commercially smoked chum salmon was tested in some Fraser River Indian villages. These experiments produced no known result.

The policy option of a complete, permanent prohibition of in-river aboriginal fishing continued to pop up long after the Canada-U.S. International Pacific Salmon Fishing Commission (later the Pacific Salmon Commission) assumed responsibility for Fraser sockeye rebuilding efforts in 1937. The mandate of the IPSFC was essentially to apply major engineering remedies to convert an already damaged but still wild river system into a new kind of river that produced the optimum yield of certain commercially important species most desired by the coast's industrial fisheries. The coast's canneries wanted sockeye, mainly, and pink salmon as well. The IPSFC, funded by both Canada and the United States, did a remarkable job of creating fishways, ladders and spawning channels aimed at restoring sockeye (the only species within the IPSFC's mandate initially) and later pink. From time to time, complete closures of the aboriginal fisheries were required to achieve these objectives, and Canadian authorities routinely implemented IPSFC requests for fishing closures aimed at rebuilding specific sockeye runs, including those on the Chilko, Nadina and Stuart river systems. Upriver tribal fisheries were routinely closed. The Ts'ilhqot'in tribal fisheries were actually closed for several years in the 1930s and 1940s as a result of federal compliance with IPSFC requests.

Certainly all salmon species contributed to the tribal fisheries prior to the Hell's Gate period, but there is persuasive evidence that chinook salmon played a much larger role than sockeye in many Fraser River tribal fisheries. The Secwepemc elders state that chinook was preferred over sockeye, and chinook comprised perhaps as much as 60 per cent of the harvest for the Stl'atl'imx people. But by the late twentieth century, chinook runs had simply disappeared from many tribal fishing areas or had been reduced to remnant runs that paled in comparison to the sockeye runs. Commercial gillnetting for chinook was closed in the Lower Fraser in the early 1980s, and what harvestable surplus of chinook remained was allocated mainly to sports fisheries. The decline in the chinook catch among Lillooet peoples was illustrated in an alarming fashion in 1987, when tribal fisheries technicians working for the Lillooet chiefs concluded that chinook bound for the entire Lillooet River system had reached the brink of extinction. An "all time low" of seventeen chinook was the sum total of the Lillooet River spawners that year. In 1988, chiefs of the Lillooet bands of Mount Currie, Port Douglas, Skookumchuck and Samahquam met and decided to forego all fishing for chinook in an effort to preserve what little remained of the species in their territories. At the time, Mount Currie Chief Leonard Andrew stated: "These are the fish which have made our cultures strong. To us, chinook is the most sacred of the salmon and our people need these fish to remain strong."

In the course of little more than a century, the salmon fisheries of the Fraser basin had gone from localized management regimes, deploying a wide variety of generally selective technologies in sustainable numbers throughout the Fraser basin and its tributaries, to an industrial, coast-based commercial fishery deploying non-selective gear indiscriminately on co-migrating stocks with differing rates of sustainable harvests. Such stock rebuilding efforts as have met with success in the latter half of the twentieth century—and without question, commercial and tribal harvests of Fraser salmon have rebounded dramatically since Hell's Gate—are almost exclusively attributable to increases in the harvestable surplus of a handful of runs of commercially important sockeye and pink salmon.

After the centre of fishing effort had long since moved out to sea, and British Columbia's tribal fisheries existed only within the confines of a food fishery that comprised about 5 per cent of the total coastwide salmon catch, the distribution of fishing effort in the Fraser's tribal fisheries, as well as the composition of species and the distribution (although certainly not the volume) of catch, still followed the general outlines of the old system up until the 1950s. In that decade, about 75 per cent of the aboriginal salmon harvest of the Fraser basin still occurred in District 1 (above North Bend, in the Fraser Canyon). By 1990, however, that upriver share had declined to less than 40 per cent, even though DFO estimated in 1989 that 82 per cent of the aboriginal population of the Fraser basin was resident above North Bend. Similarly, the aboriginal sockeye catch was still split about 50/50 between the fisheries in the Fraser mainstem and fisheries in the tributaries in the 1950s, but by 1993, about 90 per cent of the aboriginal sockeye catch was being taken from the Fraser's mainstem.

The total number of salmon caught in the tribal fisheries of the Fraser basin grew markedly between 1950 and 1990, and much of this increase is a direct result of sharp increases in the aboriginal populations of the river, coupled with the reassertion of fishing rights. It is also true that Lower Fraser native communities have been particularly persistent in maintaining their catch share through the years, and clandestine sales of food fish had encouraged relatively high fishing effort in several Lower River communities. By 1995, the total Fraser River tribal fisheries sockeye allocation was about 1 million fish.

In the main, however, aboriginal catch increases are a function of the rebuilding of stocks within the IPSFC and the Pacific Salmon Commission's stock-rebuilding mandate. Favouring sockeye, the coastal commercial fishing industry deploys gear detrimental to other, co-migrating salmon species—as does much of the Lower Fraser aboriginal fishing community, which deploys non-selective gillnets in place of long-banned traditional technologies.

While the tribal fishing effort was once widely dispersed throughout the Fraser basin and its tributaries, as the years passed the centre of effort moved out of the tributaries into the mainstem, and down the mainstem to the lower reaches of the river, following declines in species diversity. The result has been a dramatic decline in the diversity of fishing skills and in the cultural diversity of unique salmon products, processing systems and preservation methods. Fraser basin salmon harvests were once relatively spread out among unique populations of sockeye, chinook, coho, chum, pink and, to a lesser extent, steelhead. These species were also relatively spread out among the several major habitat types within the Fraser basin. By the late 1980s, tribal harvests were conducted mainly upon a few sockeye stocks. The tribal salmon fisheries of the Fraser basin had become overwhelmingly a lower mainstem sockeye fishery.

It was into this sadly diminished state of affairs that the Supreme Court of Canada arrived, on May 31, 1990, with its historic aboriginal rights ruling in the case known as Ronald Edward Sparrow v. Her Majesty the Queen.

To make a very long story short, the Sparrow case began as a routine Fisheries Act infraction involving a Musqueam man by the name of Ronald Sparrow who was found to be fishing in a food fishery with a net that was longer than the regulations allowed. The case was fought and appealed and cross-appealed to the Supreme Court of Canada. While there had been numerous court rulings affirming aboriginal fishing rights before 1990, and federal fisheries authorities had a long-established framework for coming to terms with aboriginal fishing rights, the Supreme Court's unanimous decision spelled out, for the first time, a range of implications arising from Section 35 of the Constitution, which had been entrenched in 1982. Section 35 states: "The existing aboriginal and treaty rights of the aboriginal peoples of Canada are hereby recognized and affirmed."

Among their several findings, the judges confirmed that fisheries regulations could continue to apply to aboriginal fisheries, and aboriginal rights were no more absolute than the conventional rights all Canadians enjoy. But the judges held that aboriginal rights are sui generis in nature, meaning they are unique, and rather than an "individual" right, the aboriginal right is vested communally, tribally or collectively. This one relatively minor affirmation rendered DFO's entire system for issuing individual Indian Food Fish permits unconstitutional. The judges established that tribal fisheries conducted for "food, social and ceremonial" purposes, as generally allowed under fisheries regulations, were protected by the Constitution and could not be arbitrarily obstructed. This obliged DFO to immediately abandon its practice of closing aboriginal fisheries whenever officials in charge of stock-rebuilding efforts requested closures, or whenever the commercial catch ended up being greater than anticipated. Although the judges did not find constitutional protection for First Nations in disposing of the fish they caught however they chose, the judges found that DFO was required to be "sensitive to the Aboriginal perspective itself" on what aboriginal fishing rights were all about, and that such rights should be "interpreted flexibly so as to permit their evolution over time." The ground rules the judges laid down included an admonition that tribal fisheries were not to be shut down, generally speaking, unless a valid conservation concern prevailed, and even then only as a last resort, after other fisheries had been closed. This established a constitutionally ordered priority in salmon allocations: conservation first, tribal fisheries second, commercial and sports harvests last.

Rather than attempt to establish an inventory of aboriginal fishing rights, the judges chose instead to focus on what the federal government could and could not do with regard to interfering with aboriginal fisheries. The point of the decision, in the judges' words, was to provide "a measure of control over government conduct and a strong check on legislative power."

The judges confirmed a previous decision that the Crown was bound to act in a "fiduciary," "trust-like" and "non-adversarial" manner with respect to tribal groups. These findings removed any legal justification for DFO to arbitrarily instruct aboriginal communities about how many fish they could catch, which fish they could catch, where they could fish, when they could fish or who among them could fish.

One of the key tests the judges said governments must meet in order to determine whether fishing rights had been unfairly restrained was the extent to which governments had consulted aboriginal communities and the extent to which alternatives to restrictions of aboriginal fishing had been explored. Governments were further admonished that "the honour of the Crown" itself was at stake in all these consultations. As for the general regulatory regime, the judges said regulations should not be "unreasonable," nor should they pose "undue hardship" or deny to First Nations their "preferred means" of exercising their rights.

Many speeches have been made by aboriginal leaders and by fishing industry leaders about what the Sparrow decision implied or did not imply, and both sides routinely criticize DFO for its handling of the challenge presented by Sparrow. Whatever DFO's faults or failings, it has to be said that the Supreme Court of Canada's findings left fisheries managers facing an almost untenable situation. On the one hand, the salmon fishing sequence puts saltwater commercial fisheries first in line, and those fisheries had come to account for about 90 per cent of the coastwide salmon catch, with about 5 per cent taken in sports fisheries and 5 per cent taken in tribal fisheries. On the other hand, the tribal salmon fisheries occur almost exclusively within the coast's river systems—the last stage in the harvesting sequence—and spawning, which fisheries managers struggle to maintain at levels that will ensure conservation of the resource, is the final event in the salmon's complicated life. The Sparrow decision found that Section 35 of the Constitution—which overrides the Fisheries Act, along with every other federal and provincial law in the country—had ordered a reverse system of priorities.

More importantly, DFO couldn't simply boss the Indians around any more. Without extensive consultations or pre-season agreements involving almost all the two hundred Indian bands in B.C.—almost half of which are situated within the Fraser basin—it was going to be next to impossible to plan fisheries. In this increasingly complex approach, DFO was going to be hard-pressed to make decisions unless DFO staff knew what would happen in the tribal fisheries—how many fish native people would catch, when native people would be going fishing, which fish, exactly, they intended to catch, where people would be catching the fish and how many people would be fishing.

Ottawa's response to all this perilous uncertainty was to devise the Aboriginal Fisheries Strategy, a complete, coastwide approach to the problem.

As with the Sparrow decision, many speeches have been made by both aboriginal leaders and fishing industry leaders about what the AFS implied or did not imply, and both sides routinely criticize DFO for its handling of the AFS. But for all the fuss it has caused, the AFS has never been intended for anything much more than federal Fisheries Minister John Crosbie said it was intended for, when he unveiled the strategy on June 29, 1992. In his announcement, Crosbie said this of the AFS: "It is designed to bring stability, predictability and profitability to the coastal fishery in British Columbia."

Individual food fish permits had to go, so they were replaced with communal licences, generally administered by tribal authorities, which issued "designation permits" to individual tribal members. Because of the need to consult with aboriginal peoples in order to make decisions about the commercial fisheries, the Fisheries Act was amended to allow for "aboriginal fishing agreements." To determine how many fish native people would be taking, fishing agreements specified catch ceilings in elaborate allocation regimes.

Crosbie acknowledged that substantial increases in the overall tribal catch would have to be allowed in order to ensure that DFO met its basic, minimum constitutional obligations to aboriginal fishing communities, so the AFS contained commitments to undertake a "voluntary licence retirement program" in the commercial fleet, in which unspecified millions of dollars would be spent over a seven-year period to "neutralize impacts that may result from the reallocation of salmon resources" to tribal fisheries. No formula, target or overall reallocation goal was specified, but the commercial fishing industry, through the B.C. Fisheries Commission, was assigned the task of deciding how the program would be implemented and how the money would be spent among its constituents. More than $6 million in federal funds was spent in 1992-93 to retire several commercial seiners, gillnetters and trollers, but somehow this enormous sum of money produced a mere 317,189 sockeye equivalents to be reallocated to tribal fisheries coastwide. To put that number of fish in perspective, the total commercial harvest of Fraser-bound sockeye in 1993 was about 17 million fish. By 1995, DFO had committed $35 million to the licence retirement program.

To persuade native leaders to sign AFS agreements, Ottawa offered a range of benefits, including commitments to stream-enhancement projects long sought by tribal groups; technical training in stream enumeration and other labour-intensive work for native communities struggling with outrageous unemployment rates; training and seasonal jobs in enforcement; and various fisheries co-management regimes. In its first year of operation, the AFS attracted few signatures from the Fraser basin, but since then most of the basin's tribal fisheries have slowly been brought under the AFS regime.

For years, the most obstreperous native fishing communities were those among the Nuu-chah-nulth on Vancouver Island's west coast, the Gitksan and Wet'suwet'en of the Skeena River system and the Lower Fraser tribal communities. In each case, confrontation and defiance had characterized the tribal relationship with the commercial fishing industry and federal fisheries officials, although these communities were certainly not unique in that respect. But among the Lower Fraser, the Gitksan-Wet'suwet'en and the Nuu-chah-nulth, the aboriginal leadership had long defied confinement within the nineteenth-century "food-fishing" rules. They insisted that their communities should be at liberty to sell fish from their allocations, as they had done before the Fisheries Act was extended to Canada's West Coast, and as they continued to do, when DFO wasn't watching, long after the 1880s when the food-fishing rules were first established.

Because of these entrenched positions, the AFS agreements with the Lower Fraser communities (by the second year of the AFS this included all Sto:lo bands, Musqueam, Tsawwassen and Coquitlam), the Gitksan and Wet'suwet'en bands, and two bands from the Nuu-chah-nulth contained provisions for experimental "pilot sales" regimes, so long as "existing B.C. processors" were granted rights of first refusal for all fish sold from these fisheries.

From the outset, Ottawa made it clear that these fish sales arrangements did not amount to any recognition of a constitutionally protected "aboriginal right to sell," and the tribal leadership made it plain that agreements that limited their catch and provided for a host of other controls should not be seen as any kind of surrender of their aboriginal rights. While the courts had offered conflicting versions of whether or not an aboriginal right to sell existed, Ottawa and the tribal leadership set aside their differences, agreed to disagree about points of constitutional law and got on with the job of establishing co-managed fisheries in which the sale of fish, under strict terms and conditions, would be decriminalized.

The commercial fishing industry's reaction was one of shock and outrage. For years, the industry had successfully persuaded compliant federal officials to manage salmon for its benefit. Ottawa had restrained and curtailed the tribal fisheries for a century, and had spent countless millions of dollars in various initiatives aimed at integrating native people, as individuals, into the coastal cannery-based fishing industry. In their protests against the AFS pilot sales program, industry leaders routinely cited an aboriginal commercial-fishery participation rate as high as 30 per cent, when the B.C. aboriginal population was only 3 per cent of the total population of the province: why give these people more?

Fisheries economist Marilyn James has conducted several studies analyzing federal attempts to maintain aboriginal participation in the coastal commercial fishing industry. There have been many attempts, with mixed success at best. There was the $16-million Indian Fishermen's Assistance Program from 1968 to 1979; there was the Indian Fishermen's Emergency Assistance Program, which spent about $5 million in grants and loan guarantees between 1980 and 1982. In the early 1980s, the federal government paid B.C. Packers $11.6 million for a fleet of gillnet boats the company had for years rented to aboriginal fishers on the north coast, and the boats were turned over to the fishers in a federally funded aboriginal fishing cooperative. A similar $2-million deal with the Cassiar fishing company converted fifty-two rental gillnet boats to aboriginal ownership. Still, aboriginal participation rates have risen and fallen and risen again like a roller coaster over the past half-century. Between 1946 and 1962, the number of gillnet licences held by aboriginal people dropped by 52 per cent from 1,653 to 805, but aboriginal participation in the troll fleet rose by 11 per cent, from 628 licences to 690, during the same period. The rate of native participation in salmon processing has also been subject to wild fluctuations: in 1969, Nelson Brothers Ltd. merged with B.C. Packers Ltd., which joined with the Canadian Fishing Company to take over the operations of J. H. Todd Company Ltd., and the Anglo-British Company sold its assets to the Canadian Fishing Company. Hundreds of shoreworkers' jobs normally occupied by native people were lost after three north-coast canneries shut down, along with a cannery at Steveston, at the mouth of the Fraser River. Overall aboriginal participation in the commercial salmon fleet, based on the number of vessels either owned or operated by native people, declined from 32.4 per cent of the fleet in 1946 to 15.3 per cent of the fleet in 1977.

Whatever the aboriginal participation rate was in 1992, and however stable it might have been, native communities in general, particularly communities within the coast's river systems, remained defiant. The Sto:lo maintained they should stay put on their own fishing grounds, fishing where they had always fished and selling their fish to any willing buyer, rules or no rules. It was their right, the Sto:lo said. They had always done so, through elaborate aboriginal networks that existed for centuries, networks that were expanded with the arrival of the Hudson's Bay Company, before Canada was born, forty years before British Columbia even entered Confederation. The right remained unextinguished, and after 1982 the right became protected by Section 35 of the Constitution. Or so the argument went.

Without doubt, trade in Sto:lo-harvested salmon at Fort Langley was big business, as Hudson's Bay Company records show. It began in earnest in 1830, when Sto:lo people traded salmon in the amount of 200 barrels (a barrel contained 60 to 90 salmon). The currency first in use, before English money was employed in around 1860, was usually a "made beaver," which is the value of one prime beaver skin. By 1835 the salmon traded at Fort Langley is recorded as 605 barrels, 112 half-barrels, 24 tierces, five hogsheads, and 600 pieces of dried salmon. In 1837, 450 barrels of salmon were shipped from Fort Langley to Hawaii, and production increased dramatically in the years following: 1845—800 barrels; 1846—1,600 barrels; 1847—1,385 barrels; 1849—2,610 barrels. These quantities of salmon surplus to the immediate needs of the Sto:lo people are quite staggering in a modern context: 2,610 barrels of salmon would contain at least 150,000 salmon.

In 1993, in the case of Her Majesty the Queen v. Dorothy Marie Van Der Pete, however, three of the five judges at the B.C. Court of Appeal found that while the scholarly evidence about the aboriginal salmon trade was all very interesting, such practices were at least as likely to have been rather casual, or occasional, even "opportunistic," and certainly not formalized to any degree. Since only those aboriginal customs, traditions and practices that were "integral" to aboriginal societies give rise to constitutionally protected rights, the three judges found, the Sto:lo could not show that they had rights to sell their fish any different from anybody else's rights. Two judges sided with the Sto:lo argument about rights of sale. One of the three "no" judges said he thought that while the Constitution didn't afford specific protection for commercial practices, he nonetheless thought that negotiations to allow tribal fish sales were a good idea. At the time of writing, the case had already been argued at the Supreme Court of Canada, and the judges were considering their decision.

To say that the Aboriginal Fisheries Strategy has ended up being a controversial federal policy would be to understate the case in the extreme. While many aspects of the initiative—stream clearing, spawning enumeration, hatchery jobs for native people—are without controversy, the commercial fishing industry has waged a campaign against the minority of AFS agreements with pilot sales provisions that in many ways matches the worst of the industry's previous excesses in its opposition to aboriginal fishing rights. The new campaign began with the conventional approach the industry had taken to curb the spectre of rights restoration for the previous decade: it went to court.

In 1992, the Pacific Fisherman's Alliance, the Fisheries Council of B.C. and other fishing organizations applied for an injunction in B.C. Supreme Court to suspend the AFS pilot sales programs. They lost.

The year the AFS was introduced was a year of chaos in the fishing industry. It was the low year on the Fraser River's four-year sockeye cycle, and there were fewer fish to go around than usual. The AFS also came at a time when DFO was being radically reorganized, and the fishing industry was in its usual upheaval. Salmon were exhibiting bizarre migratory behaviour, and the Pacific Salmon Treaty collapsed, prompting an old-fashioned fish war between Canada and the U.S. The headlines that dominated the 1992 season described a "biological disaster" and an "ecological catastrophe" on the Fraser, and the industry began its media campaign against the AFS pilot sales projects in earnest, blaming "massive native poaching" for destroying sockeye runs and alleging that commercial fishers were forced to stay tied to the dock while a politically correct fisheries department was turning a blind eye to it all.

Fisheries Minister John Crosbie appointed resource economist Peter Pearse and fisheries scientist Peter Larkin to investigate the allegations. Their report concluded that there was "native poaching" all right, but also that only a minority of Fraser River tribes had entered into AFS allocation agreements by then; that the Department of Fisheries and Oceans may have underestimated the sockeye catch in Fraser River tribal fisheries by more than 200,000 fish; that water temperatures associated with extremely high pre-spawn mortality rates prevailed in the Fraser River in 1992; that fish had in fact been going "missing" in the Fraser system for several years preceding 1992; and that if a variety of factors were taken into account, and if an error of a few per cent prevailed among such variables as spawning counts and estimated returns to the river, "then all the remainder is more than accounted for." Pearse and Larkin concluded that, in fact, more sockeye reached their spawning grounds in the Fraser basin in 1992 than in any other year on that cycle year in about half a century. And in the end, the native catch still came nowhere close to the pre-season aboriginal allocation, and the commercial fleet caught more Fraser sockeye in 1992 than in any year on that cycle year since the 1940s.

In 1993, the industry continued its media strategy and produced another barrage about native poaching and lawlessness, but their assault fizzled out because commercial fishermen were too busy scooping up more Fraser River sockeye than they had ever caught on any cycle since 1913. The seine boats directly owned by a single company—B.C. Packers Ltd.—caught an estimated 1.2 million Fraser sockeye in 1993, a number greater than the total 1993 sockeye allocation to all Fraser River tribes combined. For the second year in a row, the Fraser River's native communities caught far fewer fish than they were allocated in pre-season agreements, and in 1993 there was an Oka-like showdown and rail-line blockade at the Sto:lo reserve of Cheam to protest native closures while saltwater commercial fisheries proceeded.

In 1994, it was "missing fish" time again. The Pacific Salmon Treaty had collapsed again and another Canada-U.S. fish war was raging, this time euphemistically termed an "aggressive fishing strategy." Again, without doubt, there was "native poaching" in 1994. But about 2.5 million sockeye that DFO expected to see spawning in the Fraser River never made it home, and most of those fish never even made it as far as the mouth of the Fraser River. Again, Ottawa conducted an investigation. This time, it was headed up by John Fraser, a former Conservative fisheries minister. Fraser identified all the usual suspects and made a series of recommendations not entirely unlike other sensible major-overhaul recommendations made by previous commissions. Fraser's efforts were laudable, widely praised from all sides of the fishery, and they played at least a small part in focussing public attention on the alarming dysfunction in the salmon fisheries generally and in DFO particularly. But the technical investigation teams assigned by DFO and the Pacific Salmon Commission to assist Fraser and his fellow commissioners could account for no more than 800,000 of the 2.5 million fish that appeared to simply go missing in the river. The technical teams found that most of the fish losses they identified could be attributed to pre-spawn mortality in Hell's Gate, due to record low water levels and record high water temperatures. There were also simple underestimations of the legal native and non-native catch, miscounting and "poaching." As for the remaining 1.7 million fish, these were from the fabled Adams River sockeye runs, and they were nearly destroyed in 1994. It was several months after John Fraser filed his report before the Pacific Salmon Commission analyses of the Adams River disaster were in. They found that the fish had simply never made it to the Fraser River. In other words, most of the supposed 1.7 million fish were never there in the first place, and increased catch efficiency in the commercial fleet, coupled with unexpected migratory behaviour, had resulted in unintended, perfectly honest, perfectly legal commercial overfishing. It took several weeks of sales-slip analysis and catch-per-unit-of-effort reviews to come to any conclusion about how many fish the Johnstone Strait gillnetters and seiners had caught, and when all the smoke cleared, the tribal fisheries of the Fraser River basin, for the third year in a row, were the single most-blamed culprit. But for the third year in a row, the tribal fisheries' catch amounted to far fewer fish than their pre-season allocations allowed, even when "poaching" estimates were added in along with re-estimations of the legal native catch.

Nor was all well in the aboriginal communities along the Fraser, particularly the Lower Fraser communities, where AFS pilot sales programs had been introduced.

While there was a relative peace between Sto:lo communities and DFO established by the AFS, it took time for some Sto:lo fishers—many of whom had been involved in clandestine fish sales for several years—to fall into step with the new AFS regime, with its designated fish landing sites, sales-receipt paperwork and designation permits. There was some general relief within Lower Fraser aboriginal communities, and it was welcome news that families might be able to earn some extra income from sales, and to do so aboveboard, without fear of arrest. But the changes wrought by the AFS also disturbed established fishing patterns and thrust many native communities to the front lines of the underlying and persistent conflicts between aboriginal and non-aboriginal interests in Canada. The Lower Fraser's co-management structures were still being stitched together while the sockeye were passing Sto:lo fishing sites, and fisheries were already proceeding. The new regimes were developed in trial-and-error fashion, and enforcement staff management problems, clumsy bookkeeping and conflict of interest allegations didn't help matters either.

Sto:lo communities in particular (more than any other group, the Sto:lo were singled out by the commercial fishing industry and routinely castigated as "poachers" during the first three years of the AFS) were finding that for whatever its benefits, the AFS had serious drawbacks. During a series of community hearings conducted in band offices and community halls throughout the broader Sto:lo community in 1992 and 1993, involving more than twenty communities, a number of deeply felt concerns were identified. Generally, Sto:lo communities were worried that they were losing what little control they had over their own fisheries. The accustomed fishing practices they had fought for years to defend were being seriously undermined by all the new AFS regulations, and the rights they had fought so long to advance were being jeopardized by the AFS and everything that came with it.

There was a surprising degree of consensus among community members, band chiefs, councillors and tribal officials around several points. There were increasing numbers of community members joining the fishery without proper skills or knowledge. There was a growing lack of respect for traditional fishing sites. There were increasing numbers of nets in the river. And there was the potential of adverse impacts posed by the decriminalization of sales upon a variety of Sto:lo fishing traditions. Overall, people felt that "things were going too fast."

But there was also widespread enthusiasm for the prospect of restoration and conservation initiatives, a "community approach" to fishing effort, opportunities for value-added production, fishing policies that "spread around" the benefits, and enforcement of Sto:lo rules that respected Sto:lo values and involved Sto:lo people in determining "sentences" for offenders. Unfortunately, to date the AFS has been unable to deliver on any of these modest hopes.

Elsewhere in the Fraser basin, reaction to the AFS was equally mixed. The lack of a coordinated and effective all-tribes response to federal policy initiatives has confounded aboriginal leaders throughout the Fraser basin for years. Some progress appeared possible in the 1990s; there were meetings coordinated through the Interior Indian Fisheries Commission and meetings that produced protocols between Fraser River tribes, establishing various principles about mutual respect and support. At a Vancouver meeting on March 25, 1993, leaders representing all Fraser River tribal groups agreed to a coordinated approach to fishing concerns. But none of these initiatives produced much in the way of effective follow-up. There was some intertribal discussion at the First Nations-DFO Fraser River Watershed Steering Committee meetings and at some of its subcommittee meetings, but there remained the AFS "signatory" tribes in one network of meetings and "non-signatory" tribes in another. There was still no formal opportunity for the Fraser tribes to talk to one another about fishing issues outside the elaborate DFO-First Nations consultative processes established under the AFS. The federal and provincial governments had entered into agreements with most Fraser tribes to settle "land claims" by finally concluding treaties, but the B.C. Treaty Commission process provided no forum for a coordinated tribal approach to fisheries issues.

By 1995, nothing much was making sense any more. After initially estimating that a total of about 21 million sockeye would return to the

Fraser River in 1995, later forecasts put the estimated return at 10.6 million sockeye, and by mid-August the Pacific Salmon Commission, after conducting preliminary test fisheries, had downgraded its estimates to a mere 3.3 million sockeye. Boats rushed to sockeye openings only to learn that the fisheries were closed. As the weeks passed, the Pacific Salmon Commission's Fraser sockeye estimates were revised upwards, then downwards, then back up again. After the Mission hydroacoustic fish-counting facility and the Pacific Salmon Commission's test fisheries showed great numbers of sockeye in the river; openings were announced and gillnetters went fishing. But the fish weren't there. Tribal fisheries were opened, then closed, then reopened. Nlaka'pamux and Stl'atl'imx and Secwepemc people returned to their fishing stations upriver, set their nets on jim-poles and poked their dipnets deep into crevices below the white water, but their nets came up empty.

There were the usual protests against AFS fisheries. Fraser River gillnetters, among the poorest of the coast's increasingly impoverished salmon fishers, wanted the tribal fisheries shut down along with theirs. On September 15, 1995, about two hundred non-native commercial fishers towed a gillnet boat through downtown Vancouver and hung a net across a busy street, blocking traffic. Then they turned their attention to DFO's West Hastings Street office tower, where they unhinged the heavy security doors, pulled them down, broke into DFO's headquarters and occupied the place before being served with a court order to leave. The Fisheries Survival Coalition, enlivened by rhetoric about "rights based on race," rallied the public around the spectre of a federal government, acting out of some politically correct motive, taking fish away from their families and handing it over to native people.

In the real world, however, precisely the opposite was taking place, and solemn federal commitments to aboriginal communities were effectively abandoned in all but a handful of AFS agreements. One of the few tribal groups that managed to hold its ground was the Lower Fraser aboriginal fishery and its AFS pilot-sales program, which ended up only about 30,000 sockeye short of pre-season allocations. But it was a different story entirely above the Fraser Canyon. The entire allocation to the northern Shuswap and the Chilcotin River people was 100,000 sockeye, but they managed to catch fewer than 30,000 fish. The mainstem Chilcotin and southern Carrier were allocated

5,000 sockeye, but by September 19—four days after the coalition's Hastings Street protest—their total catch for the entire season was recorded as 16 fish, and that's how their sockeye season ended. The Northern Carrier and Sekani communities were allocated 65,000 sockeye, but after the runs passed their fishing stations, the catch amounted to fewer than 14,000 fish. The Stl'atl'imx were supposed to have been allowed to catch 63,000 sockeye, but they managed to catch only about half that amount. Of the 20,000 sockeye allocated to the saltwater tribal fisheries of the Pacheenaht, T'Sou-ke, Becher Bay, Equimalt, Songhees, Tseycum and Saanich communities, less than 2,000 sockeye ended up in native nets. Of the 60,000 sockeye allocated to the Island Halkomelem peoples, less than 15,000 fish were caught. At the season's end, hundreds of charges had been laid against native fishers, mainly for fishing during closures.

The initial Fraser River sockeye spawning escapement goals— the numbers of fish DFO's planners intended to make it home to spawn—were scrapped. From an initial forecast return of 10.7 million Fraser sockeye and an escapement goal of 2.9 million spawners, plans were revised downwards to reflect an actual return of only about 4.4 million Fraser sockeye. The department's hopes of 2.9 million spawners also proved to be optimistic. By November, spawning ground counts tallied only 2.3 million spawners on the grounds. And so it went.

In total, the commercial fishery catch of Fraser-bound sockeye was initially expected to have been in the several millions. By September 30, 1995, the Canada-U.S. commercial sockeye catch amounted to fewer than 1.5 million fish. The unprecedented economic consequences of the 1995 fishing season sent shock waves far and wide. Emergency measures were considered in sober discussions between bankers, fishing company presidents and senior federal and provincial officials.

In an October 19 letter to B.C. Fisheries Minister Dave Zirnhelt, Doug Kerley, the B.C. government's job protection commissioner, urged Victoria to appeal to the Canadian Bankers Association and the B.C. Central Credit Union for what he called "forbearance" to debtors throughout the fishing industry. In his overview of the industry's post-season vital signs, Kerley noted that 10 to 15 per cent of the salmon industry's big-boat fleet fared fairly well, and the coast's major fishing companies showed that "they have the resources to look after themselves." But seine crew members were facing the prospects of a winter on welfare. So were plant workers, tenderboat crews and small-boat fishers. About 5,500 workers in the industry failed to show sufficient earnings to qualify for unemployment insurance. Seine crew shares ranged from $2,000 to $3,000 after twelve weeks' work; half the troll fleet failed to show a season long enough to qualify for unemployment insurance. Fraser River gillnetters were particularly hard hit, with average gross earnings of $5,000—half a normal year's fixed costs. There were gillnet fishers who didn't even survive until the season's end, Kerley noted: "Some of this fleet did not travel to openings because they could not purchase fuel." Representatives of the chartered banks told the provincial government the situation was even bleaker among other sectors: small businesses in coastal communities, fishing lodges, tackle shops, fishing guides and hotel workers. The fishers who made money were the ones who stuck close to the Skeena and managed to catch more than the average boat's Skeena sockeye. At J. S. McMillan Fisheries' Prince Rupert operation, plant managers could boast that only 63 of their 300 workers failed to qualify for unemployment insurance. Kerley noted that, compared to the rest of the coast's salmon canneries, this amounted to a "good season."

Michael Walker, executive director of the right-wing Fraser Institute, was making the rounds with the institute's ideas for eliminating the commercial salmon-fishing industry entirely. The institute said Ottawa should relinquish the Crown's common-property title to the resource, sell it all off to the highest corporate bidder and give whatever was left to companies interested in charging fish-catching fees to tourists. The entire Strait of Georgia coho biomass, usually a resident population, had picked up and cleared off for the deep waters off Vancouver Island's west coast. None of DFO's scientists could explain why. And nobody at DFO was willing to guess how badly the Strait of Georgia coho would end up taken as bycatch in the troll fisheries on Vancouver Island's west coast, either.

Then the Adams Run almost disappeared entirely, or seemed to disappear beneath the roiling, silt-laden waters of Sand Heads, off the mouth of the Fraser River, and again, DFO couldn't say why.

It was enough to make things appear, for all intents and purposes, that Mechanism had been wrong about things all along. The old medicines couldn't cope with the new sickness. Throughout the coast, veteran DFO biologists, native and non-native fishers and commercial and sports fishers were asking themselves serious questions about what it all meant.

It must have been a lot like that in the summer of 1808, among the Nlaka'pamux people, in the days after Simon Fraser and his voyageurs—those strange, unaccounted-for beings—had left the village of Cumsheen and continued their journey downriver. Back then, it must have seemed that Coyote had been wrong about things all along.

⚓

CHAPTER VI

Brighter Water Downstream

If fishers, regulators, and coastal communities stay on the current path, marine fisheries will continue to decline, millions of people will lose their jobs, and coastal communities and low income consumers will suffer disproportionately. If instead these groups work together to improve fishery management, the oceans will be able to continue to yield fish—and economic and social benefits— for the foreseeable future.

—Lester Brown, *State of the World, 1995*

For my own part, I feel there is good reason to maintain at least a cautiously optimistic point of view. Marine organisms are renewable organisms, after all, and even when they are exploited to the point of depletion, they can usually rebound if they are permitted to do so. Moreover, many of the world's inhabitants are now considerably more disenchanted with industrialization and modernization than they were only a few decades ago, so there may be less support for development for the sake of development in the future. Other trends and movements, some of which are now gathering steam, are also cause for hope, including the trend toward economic decentralization, the 'appropriate technology' movement, and especially the worldwide environmental movement, which is garnering increasing support.

—James McGoodwin, *Crisis in the World's Fisheries*

JOHN STEVENS, THE SKIPPER OF THE GILLNET TROLLER *Wishing Well*, survived the disastrous salmon season of 1995. He says he was "one of the lucky ones." He earned enough money to pay his expenses, and he got in enough weeks of fishing time to qualify for some minimal unemployment insurance to tide him over the winter. It was just enough to keep him out of bankruptcy during the months leading up to the springtime herring fishery of 1996.

John Stevens is an optimist. He sees a future for himself and his children in the fishing industry, and his optimism is shared by hundreds of working fishers on British Columbia's coast. It is a guarded optimism, tempered by a realization of the magnitude of the changes that will be required to restore depleted fish stocks, maintain sustainable harvests and establish the kind of community stability so necessary to keep harvests sustainable.

Stevens says that despite the tragedy that underlies the past century of fisheries management of the B.C. coast, there are signs of change.

"We can turn this whole thing around, you know," Stevens says. "It's going to mean some pretty serious changes, but I think we can bring it back. We've got to stop the privatization of the fishery. We've got to have a strong federal government to deal with countries like the United States. We've got to have fisheries habitat protected by really strong legislation. But we've got to end this dictatorship from Ottawa, which is just interested in exporting raw fish through NAFTA and saying to hell with the environment. We've got to democratize the fishing industry."

Stevens's views are not the naive hopefulness of a newcomer to the industry. Now forty-six, he was already a skilled gillnetter by the time he graduated from high school.

Stevens began fishing on the Fraser River when he was eight years old, his first season as a deckhand on his grandfather's gillnet boat. Stevens's two daughters, eighteen-year-old Christine and twenty-one-year-old Dawn, are also fishers. Stevens's sister, Barbara, has worked in the industry all her life and serves as the director of the UFAWU's benefits fund. His brother Bruce works on the salmon seiner the *Ocean Venture*, and his brother Nick runs the *Emma S.*, an old troller, well known on the coast, named after the Cowichan woman who was John's great-grandmother. If the B.C. salmon fisheries could give birth, its children would be the Stevens family.

Emma was born in 1871. Sometime before the turn of the century, she married Gjan Giannaris, a Greek who jumped ship in Burrard Inlet in the days before Vancouver had a name, when the only city on the inlet was Gastown. Giannaris changed his name to Louie Peterson, an informal moniker he adopted so his friends had something they could pronounce. Unsatisfied with being confined to gillnetting with a cannery-owned licence in the Fraser River's fleet of Indians, Greeks, Finns, Japanese and Croatians, Giannaris-Peterson resolved to break out of the licencing system that held him indentured to the fishing companies, and he pleaded with his employer to sponsor his application for an independent licence. His employer was Alexander Ewen, the founding father of the West Coast salmon canning industry and owner of the massive Ewen cannery in Annacis Channel, near New Westminster. It was Ewen himself who brought the Stevens name into the West Coast fishing industry. Ewen couldn't pronounce the Greek's real name either, so he christened the immigrant John Stevens, "a good Scottish name."

Among Emma's many grandchildren was a boy named Homer. Emma taught Homer how to spear fish and dig for clams, and she instilled in him a love for fishing that was nurtured through his early teens, when he spent his summers on the Fraser River, off Ladner, fishing the same reaches of the river as his grandfather had. Homer Stevens hand-netted sockeye from his own little boat, the *Tar Box*, powered by a 3½-horsepower Palmer. He went on to a life of union organizing, and between stints in the fishery he served a term as president of the UFAWU and a term in prison as well, for a year, after defying a picketing injunction in 1967. In the midst of this active life, Homer and his wife, Grace, raised three sons and a daughter, and one of those sons was John.

"We've got to move away from the divisive kind of things that we're dealing with," John Stevens says today. "That Aboriginal Fisheries Strategy—I think it was set up to fail. Everybody's fighting each other about that and everything else, but if the communities were allowed to sit down and work things out, we'd be a lot better off. And those hatcheries—we should just let nature do its own thing. There are thousands of little creeks and streams all throughout the coast that could use a little help. Just a little help, stream clearing, whatever. You'd see a real improvement. We've just got to get people working together on these things."

From a growing number of fishers throughout the coast—commercial, sports and aboriginal—you'll hear the same kind of thing. Industrial and technological approaches to fisheries management are giving way to ideas about changing and managing human activities.

Instead of trying to manage the behaviours of fish populations, the focus is shifting towards restoring to human communities the tools to manage themselves. All the tragedies that have befallen the planet's fisheries over the past century or so—and it is only in this recent, fleeting moment in history that our fisheries have been so terribly damaged—are human tragedies. They are cultural, social, economic and political tragedies, and in each case there are a few remarkably striking similarities.

In 1994, after years of assessing fisheries management models and reviewing countless studies of fisheries management problems throughout the world, maritime anthropologist Evelyn Pinkerton and natural resources scientist Martin Weinstein collaborated on a study that attempts to identify the requisite building-blocks for ecologically and economically sustainable fisheries. In their review of fisheries that work, what became evident to Pinkerton and Weinstein is that each of these fisheries is based in an accountable human community where people have a measure of control over their own fishery as well as over the outside influences that are detrimental to it.

What Pinkerton and Weinstein show is that between the extreme of conventional, state-managed, large-scale fisheries that have predominated over the past century on Canada's West Coast, and the other extreme of privatization and the surrender of the Crown's ultimate responsibility for conservation and management, there is a middle way—a return to regional and community-based management systems and co-management systems. The best argument that can be made for such fisheries is a simple one: they work.

When the right elements are in place, regional and community-based fisheries work everywhere, with varying degrees of success. They don't work in spite of local conditions, local circumstances, local knowledge, local approaches to problems and local decisions; instead, they work precisely *because* of these things.

Community-based management systems are as diverse as the communities involved in them, so it would be impossible to simply take one model and transplant it to the Pacific Coast. But a number of general observations can be made about these fisheries. In each instance, the community is highly dependent upon the fishery; the fishery resource is highly vulnerable to non-sustainable use; the community identifies strongly with its geographic location; the community and its fishers are unwilling or unable to transfer their fishing rights to interests out of area; participants are willing to ensure equity of access and the sharing of benefits; the community is willing to exercise management rights on at least an informal basis; and participants are willing to invest resources—time, labour, energy and so on—in management as long as they have a real voice in decisions.

A key element in favour of community-based fisheries management systems—and one that may become increasingly significant as the Canadian economy disappears, to be replaced by a North American economy—is that controls and allocations designed to benefit local communities can counteract the most destabilizing influences of both NAFTA, the North America Free Trade Agreement, and GATT, the General Agreement on Tariffs and Trade. While both conventions formally prohibit "Canada-first" export restrictions on raw fish, allocations tied to local coastal communities—so long as the allocation systems don't specifically restrict exports out of Canada—may well survive challenges under international trade law. The Community Development Quotas established for local Alaskan fisheries contain built-in mechanisms that appear to have ensured local control of local fisheries resources in such a way that they cannot be construed as discriminatory export restrictions.

Local management models will work in even the most complex fisheries resources, such as highly migratory, co-mingling salmon stocks, particularly when the centre of fishing effort moves away from mixed-stock, non-selective fisheries towards more stock-specific and run-specific fisheries, using selective or live-capture technologies. Management models can take elements from modern-era co-management systems as well as from local non-aboriginal and aboriginal management systems. Just such systems are already evolving under the aegis of the Skeena Watershed Committee, a committee that involves federal and provincial authorities, coastal commercial interests, sports fishing interests and the full participation of the Gitksan and Wet'suwet'en Watershed Authority.

Pinkerton and Weinstein are not alone in their views about the dysfunction of the management models North Americans have come to rely upon, or about the political expediency that fisheries management bureaucracies have been established to serve. DFO itself, despite its near-paralysis and apparent ineffectiveness, is nonetheless an agency staffed by some of the most committed and hard-working fisheries scientists and technicians in the world. Many of these people grew up along the banks of the West Coast's rivers and were moved by their own experiences to build public service careers in fisheries science and fisheries management. Despite their often ghastly reputations, many of these biologists—and some senior DFO officials as well—have dedicated their lives to working within the system, often at great cost to their family lives and their social relationships. They have achieved innumerable small and hard-won successes in turning around what far too many observers had written off as inevitable disasters. By the 1990s, there were many veteran scientists within both DFO and the fraternity of senior salmon biologists who were making the same kind of observations as Pinkerton and Weinstein.

In 1991, Peter Larkin, one of the most respected fisheries scientists in Canada and a frequent contributor to DFO's various management system reviews, remarked: "To old-timers like me, who were weaned on the stained glass language of conservation and maximum sustained yield, management of fisheries in most parts of the world today is a morally bankrupt, unmitigated disaster ... Fisheries management is largely an exercise in documenting errors of judgment."

One of the most cogent arguments in favour of dismantling the large-scale, industrial fisheries management system that DFO has become arose from within DFO's own Program, Planning and Economics Branch in 1993. Alan Greer prefaced his discussion paper with the proviso that his proposal "does not represent DFO policy and there are no present plans to carry out such reforms." But Greer's proposal—although it can't be said to have shaken up the bureaucracy or the fishing industry at all—was in its way prophetic.

"Participants in the B.C. forest industry may have lessons to teach their neighbours in the salmon fishery," Greer observed. "Over the last decade, the reputation of the forest industry has suffered as environmental activists attacked harmful logging practices and mismanagement of the public forests. It may be just a matter of time before activists target the fishing industry for its non-selective harvesting practices. While concerns about management may be legitimate, the truth can be lost in the emotional battle for public opinion."

One truth was certainly not lost on the public, any more than it was lost on Greer: the salmon fishery was fast becoming a complete basket case. Nothing short of a devolution in management decision-making to local communities, and the establishment of cooperative management regimes and integrated fishing plans involving tribal fisheries and anglers, all engaged in increasingly stock-specific, selective fisheries, was going to save it.

By its very nature, local salmon management would require some sort of an area licencing system, Greer argued. By 1995, DFO was engaged in another major policy analysis, and some form of area licencing seemed inevitable. It had been regarded as inevitable, in fact, for years, as Peter Pearse's comments in his 1982 Royal Commission report well illustrate: "The whole coast is too large an area to regulate as a single unit, and it is treated as such only because of an accident of history."

In their ground-breaking study *Fisheries That Work—Sustainability through Community-Based Management,* Pinkerton and Weinstein demonstrate how the main "socio-political" problems that confront fisheries management around the planet are surprisingly few. Patterns repeat themselves, often proving too great for fisheries management systems to overcome and thus contributing to one collapse after another. The fundamental problems can be described as follows:

1. Undervaluing "human capital"

Fisheries science is not good science if it ignores long-standing human experience with fisheries and with fish behaviour. The observations of individual fishers and communities of fishers, along with the knowledge of local people and cultures of people, comprise a source of wealth that has been largely untapped by fisheries managers and scientists. Knowledge of this kind is every bit as important as monetary capital. It exists in the cosmology and the oral traditions and the fishing methodologies of aboriginal people, and in the life experiences of third-generation trollers who have spent their lives fishing off Vancouver Island's west coast.

It also existed in the fisheries-management intelligence of generations of Newfoundland inshore cod fishers—intelligence that was disregarded by federal fisheries managers who presided over the staggering collapse of the North Atlantic cod stocks.

The community-based inshore cod-trap and hand-line fisheries of Newfoundland were fisheries that "worked." They were efficient, low-cost fisheries that formed the basis of a modest but stable and sustainable way of life for dozens of Newfoundland villages. Just as important, however, was the knowledge of the local fishers, which could have been utilized as data in analyses of the impacts of the offshore trawl fishery.

As early as 1982, Newfoundland's inshore fishing communities were warning federal fisheries scientists that the cod stocks were in trouble, and they identified the harvests of the growing offshore fleet as the only possible culprit. In 1989, the inshore fishers were so alarmed by their observations (smaller catches, fewer juvenile fish showing up in their traps) that their union, the Newfoundland Inshore Fisheries Association, took the federal government to court. But the federal government remained convinced, based on the clean, crisp data provided by officials from the big fishing companies, that the offshore fishery was perfectly sustainable. DFO could have utilized the inshore fishers' observations to more accurately assess cod-stock structure and abundance. Instead, such local knowledge was dismissed as anecdotal and unscientific. It wasn't the kind of data that fisheries biologists have been trained to take seriously.

Fisheries management regimes are, at the end of the day, about managing people, not fish. If rank-and-file working fishers are involved in the development and design of regulations geared to achieve a certain set of results, the chances are markedly better that those results will be achieved.

2. Confusing the public interest with the interests of "powerful actors"

In Canada, as in many other jurisdictions, governments treat the interests of certain influential players as though those interests mirror or somehow represent the public interest. As often as not, the opposite is the case.

One of the most important recommendations John Fraser made in his 1994 report was his call for a "public watchdog" to oversee fisheries management and keep a close eye on the behaviour of the powerful interests that benefit from the coast's fisheries. Fraser and his panelists took particular pains to focus public attention on the importance of establishing a fisheries conservation council; Fraser described the proposal as the single most important of the thirty-five recommendations he made. At the time of this writing, however, Ottawa was still stalling on this recommendation by handing it over to a fishing industry committee to discuss—a committee of the "powerful actors" themselves.

Pinkerton and Weinstein note that despite the complete collapse of the North Atlantic cod stocks, which occurred at least in part because of the political influence of two major companies, National Sea Products and Fisheries Products International, these two companies continue as key players in the world trade of seafood products, freed of the burdens of local workers and communities throughout Newfoundland. The pattern throughout the world has been the same: "Decisions about what management risks to take may be most influenced by parties who benefit from risks the general public and the majority of fishermen would be unwilling to take. Such powerful interests may be able to avoid most of the costs of poor decisions, while the Canadian taxpayer and communities and fishermen with less influence on decisions foot the bill."

With some basic public understanding about fisheries, and greater decision-making authority vested in fishing communities, fisheries are more likely to be sustainable, and the public will not continue to bear the cost of decisions over which it has little or no real control.

3. Environmental protection costs and fish habitat losses borne mainly by fishing communities and by the general public

Public policy in Canada is moving slowly towards the "polluter-pay" principle, and governments across the country have made some slow progress in ensuring accountability for habitat and water-quality damage. But the public, fishing communities and fisheries management agencies continue to foot the bill in the majority of cases, however, which often inflates the costs associated with fisheries management by absorbing the costs associated with industrial, commercial and residential development. Conventionally, land-based industries have used rivers as "free" waste disposal areas in the same way that subdivision developers have used wetlands and creek habitat as "free" landfill sites and drainage basins. This pattern is related to a prevailing phenomenon in which powerful actors in the economy tend to pass on the environmental costs of their enterprises to already-overburdened ecosystems and taxpayers.

4. Poor compliance and inadequate enforcement

Canadians like to pride themselves on their supposed law-abiding traditions. While the American ethos is defined by life, liberty and the pursuit of happiness, Canadians have defined themselves differently, as committed instead to peace, order and good government. But as national characteristics and regional, ethnic and tribal solidarities decline around the world, old ways surrender to the new, and this is particularly true in maritime cultures that are being shaken by change. It has become a cardinal rule in fisheries management that where rules and regulations are understood and enjoy wide community support, compliance rates will be high. It is also true that there is always a small minority of fishers who will break the rules for immediate benefit, as long as the risk of apprehension is low enough.

As fish populations decline and new fishing practices and technological innovation disrupt the nature of fisheries, new rules become necessary. If violations continue, these rules often become more elaborate, more difficult to justify and less likely to be understood and supported, producing a situation requiring even more elaborate rules and heavier enforcement measures. All of this is very costly, and all of it can result in even greater rates of non-compliance. Submissions to John Fraser's 1994 public review board illustrated this pattern particularly well: the Canada-U.S. fish war (described in polite company as Canada's "aggressive fishing strategy") produced a mentality among commercial fishers that if they didn't catch the fish, American boats would. There was also widespread acceptance of the alarmist proposition that if by some chance the fish did make it past American boats, "Indian poachers" would catch them all anyway. The situation produced what Fraser called a "grab-all attitude" in the fleet.

A related problem was that, in the tribal fisheries regulated by the AFS, community opinion was rarely canvassed in the development of new regulations to meet changing circumstances. Enforcement problems became particularly acute after public anxiety about alleged native poaching ushered in ever-changing, increasingly complex rules and regulations, many of which were deeply resented in aboriginal communities. The overriding opinion about compliance with the new rules among some Sto:lo fishing families was articulated well by one Sto:lo chief who had developed over the years a particularly notorious reputation for poaching: "It doesn't matter how good we are or how well we obey the rules, they [DFO enforcement officers] are all against us anyway, so there's hardly any point. They're going to keep coming after us no matter what we do."

5. Too many big boats

This debilitating phenomenon is as acute in Canada's West Coast fisheries as in almost any fishery on the planet. But it is not simply a matter of "too many boats chasing too few fish"; in fact, it may well be that many West Coast fisheries could be made perfectly sustainable by involving a large number of small boats rather than a small number of big boats. It is not good enough to instruct the public or fishing communities to butt out of private investment decisions in boats and licences and gear, because it is perfectly legitimate for the public and for fishing communities to head off the patterns that produce uncontrollable fishing effort—patterns that establish themselves through the overcapitalization of fishing fleets. Fishers in open-access and even limited-entry fisheries find themselves forced to invest more and more money to compete with one another for dwindling harvests of fish. While it is fashionable among some economists to dismiss this phenomenon as the "tragedy of the commons," remember National Sea Products and Fisheries Products International. They're doing fine.

It may well be that in certain fisheries, under certain circumstances, a management framework that involves private property rights would be entirely appropriate, but there is little evidence to suggest that as a general rule common-property resources fare better in private hands. It is doubtful that the prairies' great herds of buffalo would have survived if they had been offered up for sale to private citizens. It is just as likely that the resource would have ended up liquidated, as it was, to make room for the production of profitable agricultural commodities more familiar to European tastes. The reality of fisheries resource "ownership" in most of the world's jurisdictions renders the tragedy theory largely inapplicable, particularly if the theory is used simply to argue against public involvement in the management of natural resources. The reality is that in Canada's fisheries, there isn't much of the "commons" left, anyway.

Throughout history, and throughout the world today, there is a staggering body of evidence to suggest that "tragedy of the commons" notion coined by ecologist Garrett Hardin in 1968 is not at all inevitable. Historian James McGoodwin observes: "The depletions of resources that have heretofore been associated with common property rights, may in fact be more a result of colonial or capitalistic policies or of industrialism or modernization." Displaced indigenous management systems resulted in few such tragedies, he says, and "in many instances such local or indigenous practices help to prevent the development of the tragedy of the commons."

Between the book ends of top-down state control and complete privatization lie a range of management options, some with spotty reputations and others with proven track records. In Canada, the major management option has been to issue limited-entry licences—a set number of licences that may be traded privately in the market-place—which allow access to certain fisheries according to a variety of rules and regulations that are determined by both the lobbying efforts of the fishing industry and the priorities established by federal managers and biologists. This option has, at best, a wildly uneven reputation.

Pinkerton and Weinstein put it this way: "Investing in a fishboat or licence is not like investing in a taxi licence." The more money invested in a fish boat, the more vulnerable the investor becomes in times of low fish abundance and fluctuations in abundance. The more vulnerable investors are, the more difficult it becomes to control their behaviour. For investors, it becomes a question of appropriate return on investment, and players become increasingly inclined to seize short-term opportunities to allow reinvestment in more profitable ventures. The more money investors have tied up in boats and gear, the more effort required to protect their investments through lobbying efforts and media manipulation. Big boats tend to have the most to lose if they can't secure enough fish to justify their costs. So the concept of involving "stakeholders" in fisheries management decisions degenerates into a simple, "rational" matter of allowing the players who have made the biggest investments (the players with the biggest "stake") to wield the greatest influence in fisheries management decisions—decisions that become increasingly less likely to serve the public interest or the long-term interests of sustainability.

6. The difficulty of defining rights of access

Throughout the world, successful fisheries management systems contain at least some means by which "outsiders" can be excluded or restrained. Even in fisheries that are merely one among many dependent upon the same highly migratory fish species, some capacity for exclusion or restraint, through interception agreements or other means, are necessary to ensure sustainability. Determining rights of access that ensure participants benefit from their contributions—be these salmonid enhancement undertakings, stream-clearing projects or meetings that must be attended—tends to ensure sustainable fisheries. "Area-based rules work best, and are most enforceable, when the costs of exclusion are not too high," Pinkerton and Weinstein observe. Unfortunately, area-based fisheries are virtually non-existent in the West Coast fisheries.

There is rarely a simple solution to the complex problem of access rights, and recent federal policy has attempted to experiment with the easiest, least disruptive solutions rather than rock the boat with more community-based options. The conversion of limited-entry licences into "individually transferable quotas"—ITQs—is currently the vogue.

ITQ management systems, generally speaking, tend to enclose open-access, common-property or similar fisheries that have become too heavily capitalized. The shares of an allowable catch in a particular fishery are guaranteed to certain participants, usually those who have demonstrated the largest and most consistent catches—which can tend to reward the very companies responsible for fleet overcapacity. Once established, ITQs can usually be sold or leased like private property.

ITQ management systems can have obvious benefits. They limit the number of vessels engaged in a fishery and eliminate the treadmill in which individual fishers must catch a higher than average share of the catch just to stay afloat. But ITQs have their drawbacks as well. They tend to "privatize" fisheries that depend on public resources; they inevitably drive up the cost of access (licence costs shoot through the roof). ITQ licences tend to be traded on the market for astronomical prices and end up being accumulated by individual companies and vessel owners in order to increase their share in a particular fishery. While many of these drawbacks can be addressed in advance by regulatory measures, such as limiting the number of ITQ licences a particular company may hold or prohibiting the "stacking" of ITQ licences on individual boats, the introduction of ITQs can be devastating to small-boat fleets and the communities they support. As the Worldwatch Institute has concluded: "ITQs ... have a certain appeal because, through transferable fishing rights, market forces can direct the allocation of resources, presumably increasing economic efficiency. ITQs have the benefit of allowing marginal shareholders to get out of the fishery with some money. The down side is that such systems allow a small number of individuals or companies to buy control over the fishery ... Small-scale fishers are too numerous—and too vital to coastal communities—to sacrifice in an effort to control overfishing."

In accordance with Pinkerton and Weinstein's observations, James McGoodwin states: "In most cases, rights of access or rights over marine resources and marine properties should be conferred on fishing communities, not on individual fishers, particularly when communities are fairly homogenous."

7.   Uncoordinated strategies and fragmented ecosystems

Throughout the vast DFO bureaucracy, as in many of the world's state-run fisheries management bureaucracies, a persistent pattern of internal conflict exists. Planners in one division may be busy pumping subsidies into improving trawl-gear effectiveness while scientists in another division are scrambling to adjust vital catch-per-unit-of-effort analytical tools—tools that will be made useless by the technological changes the subsidies produce.

Great strides have been made in coordinating terrestrial resource-management regimes in British Columbia based on an "ecosystem approach" to management. Improvements have been accomplished by setting goals for the establishment of protected areas that adequately represent separate and distinct ecosystems, and by integrating these goals with forest-management regimes that are guided by the dynamics of unique forest ecosystems. Examples of this kind of coordination have produced positive initial results across the landscape, enjoying the support of constituencies that have been engaged in long-standing conflicts—environmentalists, loggers, logging companies and rural communities, for example. Coordination of this kind is particularly necessary in the management of salmon fisheries, where hatchery planning fails to account for adverse impacts on wild salmon populations. "Ecosystem management means not focussing on a single species," Weinstein and Pinkerton point out, "but considering species interactions."

Since the rise of large-scale industrial fisheries around the planet, the implications of species interaction have been rarely considered in attempts to calculate the effects of management decisions. The consequences have been disastrous. In British Columbia, the very first appearance of large-scale natural-resource extraction on Canada's West Coast produced a most dramatic interspecies chain reaction.

Sea otters were the "resource base" for the coast's first modern-era industry, beginning in the late 1700s. In little more than a century, the coast was barren of sea otters; from the fur trade's first depletions, the chain reaction was well underway. Sea otters feed on sea urchins, and sea urchins feed on kelp; with little to regulate sea urchin populations, the coast's vast kelp forests, which once surrounded entire islands, archipelagos and rocky headlands, were greatly diminished, and so were the suites of fish populations that lived within them.

8. Inter-governmental conflict

In Canada, the various levels of government—federal, provincial and municipal—as well as various jurisdictions within government—the federal fisheries department, the federal international trade bodies—routinely exhibit ongoing conflicts among themselves, often to the detriment of fish populations and fisheries management.

The reorganization of government ministries in New Zealand presents one scenario that might provide some lessons for Canadian governments. The objective of New Zealand's ministry reorganization was to achieve more sustainable, integrated, ecosystem-based management of natural resources. Significantly, the new management model shifts much of the decision-making authority to regional boards and local councils, including Maori representatives and "stakeholder" groups.

9. Variability in supply management, product quality and product
   diversity

Economic stability in any given fishery is not a guarantee of sustainability, but without it, sustainability is made desperately difficult to achieve. Even in perfectly sustainable fisheries, economic stability is heavily influenced by a variety of factors, not the least of which are the relative market abundance of a particular fish species, which tends to determine the price paid to fishers; the freshness or quality of the fish when it is sold; and the product form the raw fish takes as it enters the market.

The nature of fisheries based on migratory stocks—no fish all year, then abundances of fish that must be delivered quickly, leaving little turnaround time before the next fishery opens or little time to waste during a particular fishery opening—tends to leave fishers in the worst possible bargaining position. Chum salmon fishing is an example. A ten-pound chum can return as little as $1.50 per fish to the harvester (although the marketing of salmon roe has helped make chum fishing a bit more profitable in recent years). These same fish, processed by smoking, and marketed and packaged carefully, could easily fetch $30 to $60 per fish. But individual fishers lack both the in-season time and the out-of-season capital to improve the profitability of their fisheries.

Fishing cooperatives have produced some successes, and it seems clear that the cooperative efforts of fishing communities hold the promise of stabilizing profitability and increasing the likelihood of sustainability in the process.

On Canada's West Coast, these nine problems exist, in varying degrees, in all our fisheries. There are ways out of this mess, however. There is no magic formula, and there is no happily-ever-after, and there can be no single approach that will best suit every fishery and every community. But there are a range of options, some arising from ancient systems with proven track records, others from more recent innovations that have demonstrated tremendous potential. In B.C., there are three issues of particular concern that will have to be addressed before charting a course in the direction of sustainable fisheries: the often-conflicting resource demands of native and non-native fishers; the challenge of maintaining vital fisheries management functions during an age in which public resources are disappearing and government is downsizing; and the search for alternatives to privatization of fisheries as a consequence of international trade arrangements.

With regard to aboriginal fishing rights, the overwhelming public perception—and the overwhelming policy response—is that these rights are merely guarantees of amounts of food for native people. Historically, the central conflict has been between aboriginal communities that insist upon maintaining an independent, evolving economic relationship with their traditional resources and the coastal commercial fishery complex, which has sought to eliminate aboriginal fisheries entirely, or at least restrict them to all-you-can-eat affairs. As a result, public policy has been geared to what the courts have had to say about the constitutionally protected rights of aboriginal people to continue to fish for food. DFO routinely attempts to discharge its constitutional obligations by issuing permits to seine boats to harvest pre-set amounts of "food fish" for delivery to aboriginal communities, as though aboriginal rights can somehow be honoured by providing food-fish as a sort of welfare supplement. This approach is grotesque, but at least it is consistent with federal policy: DFO has tended to manage "Indians" the same way it has managed "fish."

DFO has managed "salmon" and "salmon stocks" instead of allowing the management of specific fisheries upon specific, distinct populations of salmon; the department manages "Indians" by providing them with "salmon" regardless of the specific fish populations the aboriginal communities traditionally harvested, and usually without regard to the communities' specific customs, conventions or fishing patterns. This may well be defended as a rational and necessary policy within the complex coastal fisheries management and the "fully subscribed" salmon resources ostensibly managed by DFO. But it has little to do with aboriginal fishing rights.

Whatever the issue at bar, the courts have made plain in case after case, at the Supreme Court of Canada and at the B.C. Court of Appeal, that "Indians" don't have "fishing rights." Specific aboriginal peoples have specific fishing rights, and those rights arise from their specific fishing customs, fishing traditions and fishing practices. These rights cannot be determined in a vacuum; rather, they must be determined by considering the customs, traditions and practices of specific communities. If these customs, traditions and practices can be said to have been integral to the aboriginal society, then there are rights involved, and these rights enjoy the full protection of the Constitution. Except for a justification as important as conservation, DFO's policies and regulations must conform with these rights. These rights may be exercised in a contemporary manner, and fisheries managers are obliged to be "sensitive to the Aboriginal perspective itself" on what these rights mean.

Fishing rights are as complicated as the migratory behaviours of the salmon populations upon which aboriginal fisheries were directed down through the generations. The continuing decline of fisheries leaves native and non-native fishing communities sadly diminished as the "human capital" of this experience—the vast body of local knowledge, fishing technique, methodology and management intelligence—is liquidated. As long as this trend continues, our options are diminished, and the options we leave to our children, whether native or non-native, will be diminished as well.

The federal government's response to the obligations the Supreme Court of Canada spelled out in the 1990 Sparrow decision has been a story of lost opportunity. The decision could have been interpreted— as subsequent senior court judgements have made abundantly clear— to provide a constitutional guarantee for cultural diversity in tribal fisheries and genetic diversity in salmon populations. This would go a long way to ensure long-term benefits for *all* resource users. Instead, the Aboriginal Fisheries Strategy was little more than an elaborate, bureaucratic policy designed to ensure that the status quo could be maintained. It was designed to shore up the dam Coyote dismantled at the beginning of the world. For all the promise of AFS pilot sales programs in the Fraser River, the initiative has tended to simply convert tribal fisheries into what appear not altogether unlike limited-entry commercial fisheries. Regardless of how the courts will ultimately rule on the commercial aspects of aboriginal fishing rights, one has to wonder how aboriginal people might enjoy a right to sell fish— whether or not this is a merely statutory or a full constitutional right— if there are no fish left anyway. The road the AFS did not take was the more difficult work of restoring to the tribal communities greater control over their fisheries, allowing them to decide for themselves how to restore the economic relationship they have always maintained with the resource in cooperation with neighbouring non-native fishers.

As for government downsizing and the increased scarcity of government resources, it has reached the point that most of the costs associated with West Coast fisheries management are now probably a result of the fisheries' increased complexity and overcapitalization. It is quite likely that in the case of many salmon fishery openings, particularly for low-priced species such as pink salmon or chum salmon, the costs associated with the decision-making process preceding the fishery—never mind the costs of enforcement, monitoring and so on—far exceed the total landed value of the fish harvested. Large-scale fisheries management is not going to get much cheaper as the years pass. Privatizing the fisheries might well eliminate these immediate management costs, but there is no evidence that such fisheries would be sustainable; there is also little evidence that these fisheries would produce any demonstrable community benefits, and it is unlikely that most Canadians would contemplate such a surrender.

To continue to manage large-scale industrial fisheries under conventional, direct federal authority, however, is to live within something of a fiction. The North American Free Trade Agreement has seen to that, as did the General Agreement on Tariffs and Trade, both of which deny Canada the right to restrict the movement of unprocessed fish out of the country. So it is going to be necessary to consider what the alternatives might be. As an alternative to ITQs, Community Development Quotas (CDQs) could begin to capture public benefit from fisheries, even those directed upon the offshore, deepwater species currently targeted by the West Coast's trawl fisheries. With community-based fisheries that place a higher priority on sustainability, fisheries for groundfish and other species targeted by trawlers could begin to employ selective technologies, new technologies could be tested, and jobs could be maintained and likely even created. The relatively small work force involved in the trawl fishery could easily survive a dramatic shift away from the large-scale, company-boat fishery that had evolved by the 1990s. Despite expensive subsidies in trawl-gear selectivity research, mesh-size regulations and other experiments, the West Coast's trawl fishery, by the 1990s, could still not withstand the simplest, most basic tests of cautious or conservative management. There was little that could justify the fishery, by any analysis.

But this does not mean that a protein-hungry world must go without the fish the trawlers catch, or that fishers must look to another line of work, or that community benefits and national benefits must be foregone, sacrificed on the altar of sustainability. The opposite is the case. In many instances, there are other, more labour-intensive gear types that can catch the trawlers' target species without strip-mining the sea bottom, and they produce higher-quality fish in the process. Among these technologies are jigging, longlining, trap fishing and pot fishing. As Greenpeace Canada's 1995 study of the trawl fisheries concludes: "These fishing methods are easier to regulate, generate higher unit prices due to better fish quality and recovery, employ more fishers, and if properly managed, will offer far less risk to the fish stocks than does trawling."

It is not a matter of replacing one group of fishers with another. It is rather a matter of replacing short-term profits with long-term community benefits, and there is no reason why the existing skippers and crew members in the trawl fleet should not be assured a place on the ground floor of a reconstructed, sustainable fishery. Fishers are an innovative and imaginative bunch. The chances are good that the trawl fishers themselves, working with longline fishers and trap fishers—rather than just with another committee of DFO department heads and biologists—would prove themselves perfectly capable of coming up with more appropriate technologies.

Just such an example of that kind of necessary innovation is coming from the gillnet fishers on the Lower Fraser River, where one of the most serious bycatch problems in the West Coast fishing industry occurs during the autumn gillnet fisheries for chum salmon. When gillnetters set their nets for chum, they also intercept vulnerable and endangered runs of steelhead and chinook salmon.

Mark Petrunia is a long-time Fraser gillnetter. He moors his boat, the *Escorial II*, to his houseboat in Annacis Channel, just downstream from New Westminster, where he lives with his wife, Mary, and their kids, Matthew, Willie, Chrissie and Samantha. For years, Petrunia has watched his fishing time shrink and his income fall. He still keeps his chinook nets, living in the hope that the Fraser River's gillnet fisheries for chinook salmon, shut down in the late 1970s, will reopen some day. With the passing years, sportsfishers have been putting increasing pressure on DFO to shut down the Fraser gillnetters' chum fisheries because of their bycatch of autumn-running chinook and steelhead. Petrunia said he saw the writing on the wall, and took it upon himself to act.

"We have to do something, or else we're not going to be able to catch chum because of the steelhead and all that," Petrunia says. "So I figured I had to try something different."

What Petrunia reckoned was that he might be able to catch a chum salmon—a fish that usually requires a gillnet with a seven-inch mesh size—with an oolichan net. Gillnets for oolichan, the tiny smelt-like fish that return to the Fraser to spawn in the spring, are small-mesh nets, with openings of an inch and three-eighths. Petrunia applied to DFO for a scientific permit to allow him to conduct his experiments. Officials were a bit confused at first, but once Petrunia explained that an oolichan net might be capable of catching a big chum salmon by entangling the chum's distinctive, protruding teeth, rather than by the chum's gills, Petrunia's permit was approved.

In the fall months of 1995, Mark Petrunia headed out to his favourite spot in the river, rolled his oolichan net from the stern drum of the *Escorial II* and waited. He remembers watching his corkline bouncing wildly. There wasn't going to be a problem getting in the path of homeward-migrating salmon, that was for certain. Snaring the fish was the problem. During four half-hour fisheries, Petrunia managed to catch five chum salmon, no steelhead and no chinook salmon. Now he's hoping to find a net with a three-inch mesh, a mesh size he says should be just enough to ensnare a chum salmon's protruding snout as effectively as a regular-sized chum net would ensnare a fish by its gills. But Petrunia's "teeth net" would allow passing steelhead and chinook salmon to bounce off and swim away. Those non-target fish that did get snared could be released live, because unlike a gillnet, a "teeth net" doesn't drown the fish.

"You could be as selective as you wanted," Petrunia says. "That's the kind of thing we're going to have to start to do. The whole industry.

"The fishing industry has got to totally change. If there's bad fisheries, and they don't change, you get rid of it. Some of these fisheries, like the seiners in Johnstone Strait, we've got big problems there. That's a fact. What I'm trying is something I'd like to prove. I'd like to prove it could work. Maybe it would take a lot longer to catch the same number of fish, but so what?"

Another veteran of the Fraser River gillnet fleet, Barry Manuck, had been thinking about the same kind of thing for years. Like Petrunia, Manuck grew up on the river. He and his wife, Nancy, had also hoped they could establish a place in the fishery for their children, Kevin, Shane, Steven and Jeremy. But Manuck was tired of roaming the coast in the family gillnetter, the *Panther*, looking for fishing openings. He was watching with alarm as it became increasingly difficult to make a living on the Fraser, and he felt helpless as the river's fishing communities were being divided into opposing, hostile camps of native and non-native fishers. Around the time Petrunia was applying for a scientific permit to experiment with a small-mesh net for chum salmon, Manuck began talking to fishers from the Katzie First Nation, at Pitt Meadows, about teaming up on an even more ambitious plan.

Manuck proposed a live-capture, beach-seine fishery for chum salmon. He brought together several non-native gillnetters who agreed to abstain from gillnetting chum salmon in the Fraser for the season. Katzie band councillor Rick Bailey agreed to work with Manuck, and several Katzie fishers agreed that they would abstain from any gillnet fishing in the aboriginal fishery for chum salmon. Instead, the two groups of fishers would join in a beach-seine fishery—one end of a heavy-webbed net would be anchored to a beach, with the other end pulled out into the river and drawn in an arc, eventually encircling the homeward-swimming fish. The net would be pulled close in to shore, with non-target fish allowed to escape alive. A formal proposal went to DFO for a 30,000-chum allocation, with a third going to the natives for food, social, ceremonial or sale and a third going to the non-natives for their own fridges, smokehouses or sale. The final third would be sold, with the proceeds "put back into the resource" by employing participants from both communities in winter enhancement projects like stream clearing and other labour-intensive activities.

At first, the plan drew threats of boycotts and blockades from some non-native gillnetters. DFO was concerned about protests, but mostly, the department's Fraser River division officials were reluctant to attempt a "new fishery" on such short notice, even if it was experimental. Eventually, DFO agreed to issue Manuck and Bailey a scientific permit for 3,000 chum—one-tenth the proposed allocation—with proceeds to be spent on an agreed-upon enhancement project. Then the rains came, and with the high winter tides in the river, it was impossible to find a suitable beach, and by the time everything was in place the fish had already passed. But the gillnetters and the Katzie fishers made a couple of trial runs anyway, just to see what it was like.

"It worked fine. If there was fish there, we could have caught them," Manuck says with a laugh. "But we're not going to give up on this kind of idea. I think this kind of thing is the way to go, and I guess we got people thinking, because we've had guys from all over asking about it, from as far away as Steveston, and they want in on it."

By 1995, there were fisheries in B.C. that were working, and they held the potential of providing lessons for fisheries throughout the coast. One of the most encouraging management models, emerging on the Skeena River, had gone a great distance in addressing the kinds of problems that existed in the Fraser River fisheries. The Skeena management model was one of the case studies examined by Pinkerton and Weinstein.

In the decades that passed after Hans Helgeson was honoured by the coastal canneries for the part he played in dismantling the Skeena's upriver tribal fisheries, several distinct salmon runs harvested by the Gitksan and Wet'suwet'en people had been driven to extinction, and many of the runs that remained were severely diminished. The

Skeena's sockeye runs alone were made up of at least fifty-seven separate spawning populations, but this diversity too was radically diminished. Over the years, relations between the Skeena's upriver tribal fisheries and the commercial fisheries of the coast became strained to the point of violence. Confrontations between Gitksan fishers and DFO officers were common. There were court battles over the Gitksans' proposed rights to sell fish and increasing animosity between recreational fishers and tribal fisheries. There was growing federal-provincial tension over the situation; the provincial government is responsible for steelhead conservation, but senior federal fisheries officials continued to condone mixed-stock gillnet fisheries in the saltwater and at the Skeena mouth that were depleting steelhead runs. By the 1980s, some recreational fisheries groups were calling for the elimination of commercial net fisheries altogether in order to save the Skeena's steelhead and coho populations.

Whatever its faults, DFO had implemented successful stock-rebuilding measures for Babine Lake sockeye—the fish Helgeson described in smokehouses covering acres of ground back at the turn of the century. The construction of spawning channels on the Pinkut and Fulton rivers produced such an abundance of sockeye returning to those rivers (runs that could easily sustain harvest rates of up to 80 per cent) that restraints on the mixed-stock fisheries at the Skeena mouth became necessary to protect co-migrating salmon runs that could not sustain such high rates of harvest. Even the most minor commercial fishery restraints routinely meant a wasteful culling of sockeye runs as they entered Babine Lake, to ensure that the Pinkut and Fulton spawning channels were not plugged beyond capacity with spawning fish. Time and again, proposals by the Gitksan and Wet'suwet'en people to partially revive their upriver fisheries on Babine-bound sockeye, allowing some commercial sale, were met with emotional opposition by coastal fishing interests swept up in the rhetoric of "rights based on race." By 1991, the situation looked about as hopeless as any fisheries controversy on the coast.

Following interventions by the Steelhead Society, however, the tribal, recreational and commercial fishers started to talk to one another about the problem. DFO, which was facing mounting pressure to take action to protect the Skeena's "weak" stocks, warned commercial fishing interests that unless they came up with solutions, solutions would be imposed on them. By 1992, there was a small patch of common ground in a memorandum of understanding that united the warring parties around certain principles, not the least of which was an agreement that the problems required local solutions that accommodated all the Skeena's communities.

By 1994, the Skeena Watershed Committee had been formed, federal funds were committed to conducting extensive research on steelhead and coho stocks, a pilot-sales project was in place in the river's tribal fisheries, and commercial fishers were participating in the development of more precise fishing plans that reduced the bycatch of non-target salmon species. At the time of this writing, the Skeena Watershed Committee was working well. The harvest rate on Skeena steelhead was down to between 21 and 22 per cent, almost on target. Each of the three sectors—commercial, aboriginal and recreational— had opportunities to review each other's fishing and enforcement plans. Cooperative enhancement and habitat restoration planning was underway, and confrontation and conflict had become the exception to the rule. The Skeena Watershed Committee was performing management functions on an informal basis, including stock assessment, resource use coordination and policy development. Tsimshian tribal fishers were deploying selective beach seines in the lower reaches of the Skeena, and Gitksan and Wet'suwet'en community leaders took the opportunity of their pilot sales project—a much smaller-scale project than the Lower Fraser's—to reestablish their selective fisheries, experimenting with "modern" materials. In place of the sealskin floats and cedar withes used in ancient fish traps, the tribal fisheries technicians used empty propane tanks and alumimum rod. The experiment was a bit of a bust, but it was an honest effort, and they learned some things they had forgotten from the days before Hans Helgeson journeyed upriver to tear down their "barricades." Tribal fisheries planners turned to another live-capture experiment with fish wheels, devices powered by the current of the river that look much like ferris wheels. The fish wheels worked.

Such cooperative, community-driven approaches have shown success in fisheries that exhibit even greater complexity than the Skeena's salmon fisheries.

In Alaska, following a period of dangerous declines, the state's salmon fisheries were producing staggering sockeye catches by the 1990s. In 1994, Alaska's salmon catch reached an unprecedented 194 million salmon. A unique feature of Alaska's salmon fisheries is a system of area licencing, and the state has established a system of regional salmon-enhancement institutions. One of Alaska's major salmon-producing rivers is the Kuskowim, a 1,600-kilometre river that produces chinook, coho and chum salmon and empties into the Bering Sea. There are twenty-one communities strung out along the Kuskowim River, and the players include state fish and wildlife authorities, native Yup'ik fishing communities, fish processors, commercial harvesters, subsistence fishers and recreational anglers. Unlike in B.C., the lines between these sectors are not so clearly demarcated. In Alaska, "white" people enjoy "food fishing" rights—which enjoy priority over other fisheries—and the same white people also participate in the commercial fisheries. Similarly, Yup'ik people persist in their traditional fisheries and also enjoy the same "food fishing" rights at the same time they are involved in the river's commercial fisheries.

This turnaround in the Alaska salmon fishery occurred only recently. In the 1980s, after disturbing trends in chinook-spawner escapement data and rising commercial catches of chinook, Alaska's fish and game department was seriously considering a shutdown of the Kuskowim's commercial chinook fishery.

The tasks of conducting run-strength assessments in the silty waters of the Kuskowim, the distances involved between fishing areas and communities, and the diversity of the players involved all combined to make fisheries management on the Kuskowim every bit as complex as fisheries management on the Skeena, if not more so. As in the case of the Skeena, however, local communities along the Kuskowim were given an opportunity to grapple with these problems cooperatively. As is not the case on the Skeena, cost-recovery in fisheries management has been a priority. Fish processors in Alaska played a key role in paying for the costs of run-strength assessments, and state authorities soon agreed that it was a lot more expensive to maintain a ninety-day "scientific" fish-counting system than it was to simply pay a local, well-trusted Yup'ik fisherman who lived at Eek to go fishing where he lived and call in his results over a crackling radio-telephone. The subsistence fisheries were used as run-strength estimate systems, and the success of Yup'ik fishers near the mouth of the Kuskowim proved a more accurate gauge of abundance than elaborate computer models based on test-fishing results tabulated by government scientists. At regular meetings of the Kuskowim Working Group, time was set aside on the agenda for a "traditional knowledge report" that was given as much weight as the standard scientific data usually discussed at such meetings. While the Alaska department of fish and game still maintains the ultimate responsibility for fisheries management and conservation, the functions of management and conservation have been largely taken up by the Kuskowim Working Group. The working group has no formal, legal authority. It doesn't need any.

On the far side of the Pacific Ocean, one of the planet's most complex fisheries management regimes, and one of the planet's most productive fishing areas, shares many of the characteristics that produced the Skeena and Kuskowim regimes—community-driven management, the presence of accountable fishing communities with a high dependence on the fishery, and an unwillingness to alienate the resource from the community. But unlike the Skeena and the Kuskowim fisheries, the Japanese inshore cooperative fisheries produce volumes of fish in the range of three million tons annually, about ten times the volume of the fish produced by all of British Columbia's fisheries combined. The Japanese inshore fishery produces about one-third the volume of all of Japan's fishing fleets combined, and about half the value of Japan's entire commercial fishing industry.

After American and European merchants forced Japan to open its markets to the rest of the world in the nineteenth century, the Japanese approach to the challenge of industrialization was to search the world for the most effective economic and governmental models and implement them at home. Delegations dispatched throughout the planet came back with detailed reports. Japan adopted many of Germany's forms of military organization, replicated aspects of Britain's education system and took bits and pieces of other management systems from other nation states. After some disastrous attempts at replicating other nations' fisheries management models—such as central-state management of the type that has governed most of Canada's fisheries— the Japanese decided that no system provided methods any better than their own. By 1901, national legislation vested management rights and responsibilities in local cooperative associations, which followed the lines of the feudal-period fishing village guilds. The system ensured that only local residents would participate in locally managed fisheries and that the benefits would generally accrue to the local communities.

Japan's inshore fisheries have suffered from many of the same syndromes that have plagued Canada's fisheries—technological advances, inequity, the advantages of capital over labour and the emergence of

wealthy processors and middlemen. In the late 1940s, however, much
of this dysfunction was resolved by legislation that further entrenched
local controls and curtailed "absentee" players. In this respect, Japan's
inshore fisheries have already faced the same upheavals Canada's
Pacific fisheries are now facing. And for the Japanese, the way out of
these crises has been to move in the direction opposite to the one
Canada has taken down through the years. Formally, Japanese inshore
fishing rights are held by prefectural governments, each of which is
much like what the Central Coast Regional District might look like if
the B.C. coast's regional districts enjoyed greater authority and natural-resource jurisdiction. The Japanese inshore cooperatives' fishing
rights are renewed every ten years.

   If there is any argument against locally based fisheries management
on the grounds that such regimes are too small in such a big world, or
that enforcement problems would be extreme, or that there are already
too many boats chasing too few fish, Japan's inshore fisheries prove the
opposite. (There is an irony in all of this, of course. Japan has
developed a notorious reputation in some high-seas fisheries, including
the mid-Pacific squid driftnet fishery. But it is worth noting that one of
the main reasons Japanese driftnet ships were out in the middle of the
Pacific Ocean in the first place is that the inshore fishing communities
had successfully banned them from Japan's two-hundred-mile zone.)
Japan's inshore fisheries appear to be perfectly sustainable. Many have
continued for several centuries, and fish production remains relatively
stable from year to year. There are 2,127 fisheries cooperatives involved
in the management system, each with a membership average of 250
fishers. The cooperative membership exceeds 500,000 people. Each
household typically owns one or two small boats, with motors limited
to thirty horsepower. A wide variety of gears are deployed. Fisheries
are sometimes mixed-stock, sometimes not, and there is a dizzying
variety of fish species involved in these regimes.

   Each cooperative is responsible for establishing its own fishing plan,
which must be approved by the cooperative's membership as well as by
a government regional fisheries commission. This does not guarantee
that all of the people will be happy all of the time, but the regime does
ensure that those with a direct stake in the fishery have opportunities
to sort out problems with their fisheries and to have their positions
heard. A 1989 study of the system provides this useful illustration of
how conflicts are resolved:

"Meetings of groups or squads are often loud and noisy. Fishermen speak out frankly, and are not intimidated by the leaders. Emotional outbursts are common, and the meetings often break up temporarily into small groups of heated debate. Ideally a meeting should produce a consensus that takes into account the interests of all the parties. To close a meeting by disposing of objections through majority rule would be normally unacceptable.

"Once a definition has been reached and the meeting adjourned, the men send out for rice wine and dried fish, and a small party begins. During this time the men have an opportunity to assuage feelings hurt and egos bruised in arguments or compromises."

With the exception of the rice wine and the dried fish, the above description might well fit several emerging, local-management B.C. initiatives.

By 1995, on Vancouver Island's West Coast, a local, grassroots organization known as the West Coast Sustainability Association (WCSA) had brought together interests that have conventionally remained locked in constant conflict. Commercial salmon fishers and tribal leaders, already concerned with severe stock depletions and habitat loss in the area, were facing the prospect of being shut out of the industry by coastwide allocation decisions that would have centred fisheries too far north for many small-boat trollers to safely travel. While there was no shortage of tension and mistrust aggravated by the Aboriginal Fisheries Strategy, the two sides decided that things were getting so desperate it made better sense to keep the lines of communication open rather than retreat into warring camps.

For years, native and non-native commercial fishers had already been working together on cooperative projects with tribal fisheries authorities, local sportsfishing groups and federal fisheries officials. These projects were mainly small-scale efforts, most notably the activities of the Thornton Creek Enhancement Society. Commercial trollers involved with volunteer hatchery work and brood-stock collection were also beginning to discuss joint initiatives with native groups such as the Tla-oqui-aht First Nation, which had been working mainly on its own to rehabilitate Kennedy River chinook stocks.

Kennedy Lake sockeye had been severely depleted over the years. Sockeye were returning to the lake in numbers up to 160,000 in the 1960s, but these runs had declined to levels that ranged from 7,000 to 60,000 by the 1970s. By 1982, the Tla-oqui-aht people had voluntarily

abandoned their food fishing rights on Kennedy Lake sockeye. Working with the Portland-based Ecotrust International (an environmental
organization that was becoming increasingly involved in the development of economic alternatives to clearcutting in the Clayoquot Sound
area), the Tla-oqui-aht and the Nuu-chah-nulth Tribal Council began
to formulate restoration strategies for Kennedy Lake sockeye. Out of
these early plans emerged the Kennedy Lake Salmonid Technical
Working Group, which brought together the tribal communities, local
native and non-native commercial fishers, area environmentalists, the
B.C. forests ministry, DFO officials and MacMillan Bloedel, the major
forest tenure-holder in the area. Working to restore Kennedy Lake
sockeye, participants were aided by a key strength of the Kennedy Lake
initiative—a reliance on both conventional science and local knowledge. While the working group had no formal management authority,
its recommendations concerning data collection and analysis, stock
assessment, conflict resolution, habitat restoration and even harvest
levels carried a tremendous amount of weight, precisely because all the
key local interests were involved.

All these activities laid the groundwork for the West Coast Sustainability Assocation, which likewise has no formal authority but also
possesses a high degree of moral authority in fisheries-related decision-
making. Based on an equal share of power between the native and
non-native communities of Vancouver Island's west coast, including
recreational fishers and civic officials, the WCSA, by early 1996, was
turning its attention to one of the most difficult and controversial
fisheries issues in the North Pacific—the offshore trawl fishery.

At the time of this writing, the WCSA was laying the groundwork for
a complete fisheries management regime, in cooperation with the local
regional district, local fish processors and local Nuu-chah-nulth villages, to take over the federal government's management functions in
the hake trawl fishery. Their plan would look a lot different from the
regime that exists under DFO's aegis.

"The community believes it is inappropriate for its future economic
dependence on fisheries to be determined by a government agency and
its advisors without any representation or consent to the manner in
which the fisheries adjacent to the community are to be used," the
WCSA's proposal asserts.

Some of the key problems the WCSA proposal identified with the
hake trawl fishery were: a complete absence of any certainty for the two

hundred local processing jobs associated with hake catches; an absence of any certainty that long-term infrastructure costs borne by the community, such as roads, sewers and other services, could be recovered in light of the highly mobile fleet's focus on short-term profits; the likelihood that any minor disadvantage to the fleet in landing hake in domestic ports would result in the fleet landing its catch in U.S. ports or off-loading to foreign factory ships; inadequate fleet monitoring and the fleet's inherent tendency towards overfishing; and DFO's emphasis on the immediate interests of a highly mobile fleet rather than on long-term fisheries sustainability and community stability.

The WCSA proposal's main elements were: the development of a cooperative arrangement with the federal government in which a regional fisheries management board, representing all the major interests in the fishery, would set harvest levels and organize fisheries within federal quotas; technicians and at-sea observers who would monitor bycatch and other environmental impacts of trawling, maintain detailed landings data and perform other necessary duties; the recovery of fisheries management costs through the sale of hake harvested in a special revenue-generating quota within the overall hake allocation; a specialized, hake-only licencing system that would exclude all other trawl vessels from participation, ensuring against overcapitalization; and a range of checks and balances that would be put in place to directly link the long-term health of the local hake resource with the long-term economic stability of local communities.

The proposal notes: "As has historically occurred in most industrial fisheries, overfishing and stock depletion of one species of fish has resulted in fleets moving to more distant fishing grounds and to alternative species. The result is coastal communities with displaced labour opportunities and depleted fish resources."

Ucluelet troller Dan Edwards, a director of the West Coast Sustainability Association, says he's convinced that only in regional, multi-party fisheries management does sustainability have any hope of taking root.

Edwards is a third-generation West Coast troller. His grandfather, Ernie Edwards, emigrated from Newfoundland in the 1920s and was among the first of the non-native trollers on Vancouver Island's stormy outer coast. Ernie Edwards's son Glen spent a life in the fishery, and Glen's son Dan spent his first season aboard his father's boat at the age of seven. Dan Edwards trolled for salmon, worked in logging camps and fished for prawn, halibut and crab, and after a brief stint in university, he returned to trolling. Edwards's twenty-one-year-old son Ryan is a troller, and his nineteen-year-old daughter Danielle is at university, studying environmental science and sociology.

"I'm a fisherman, but fishermen shouldn't be the only part of management, you know," Edwards says. "You've got to have a mix of values in it. DFO is supposed to represent those values, but it can't. That whole central-management thing has had its day."

Edwards was in on the ground floor of the initiatives that paved the way for the WCSA, and he cautions against viewing the cooperative initiatives at the association's inception in too rosy a light.

"All those tensions are still out there in the community," he says. "You got hard-core people, and some of the natives are worried about this kind of thing, just like some of the non-natives are worried. Some of the DFO people are really worried. There are a lot of people who say, 'We don't want to do this, we'll lose out somehow.' In the industry, you've got lobbying pressures that are really hurting the ability of people at the regional level to be effective. When we say that we've got to get all these people together, we're even creating tensions just by saying that. But we've got to resolve these tensions—in the community. We've got to be a spur in the side of the people who think the status quo is just fine.

"So the way I see it is, you got to heat it up. Get everybody in the room. Battle it out. Eventually, we'll get there."

In the late 1980s, the same kinds of challenges that the West Coast Sustainability Association was trying to address—the industry's lack of accountability, overcapacity and chronic overfishing—were among the difficulties facing both native and non-native clam harvesters on the Sunshine Coast and the Malaspina peninsula. The lack of adequate science, along with fisheries managers' lack of familiarity with local knowledge, also presented serious problems. The emerging solutions to these problems provide a glimpse of the kind of sustainability and community stability envisioned by the West Coast Sustainability Association.

For thousands of years, the local Sliammon, Sechelt and Klahoose peoples maintained intensive shellfish harvests in their traditional territories, regulated by local customary law. After federal authority in fisheries was asserted, the management of shellfish harvesting in these areas became a relatively simple affair. At first, no licences were required to dig clams. Then clam-digging licences were issued to all comers. Later, a single management unit was established—"Area C," which takes in a vast stretch of the mainland coast from the Canada-U.S. border to the straits and islands of Desolation Sound, north of Lund.

By the early 1960s, commercial shellfish harvesting south of Howe Sound was closed because of pollution. But shellfish resources north of Howe Sound, formerly dominated by butter clams, became more diverse, as a consequence of the introduction of manilla clams. Because of local environmental conditions, harvesting effort in the vast management unit that Area C comprises tends to be concentrated in a few favoured beaches: Savary Island, for instance, accounted for about 50 per cent of the entire Area C manilla clam harvest by the early 1980s, when the market for manilla clams was booming. But the number of clam diggers in the area was getting out of control. In response, DFO established fairly arbitrary quotas on the number of clams that could be harvested in the area and implemented a "closed" season during the summer months throughout Area C. These measures proved wholly inadequate. The harvesting of undersized clams was routine. Savary Island's beaches had become overcrowded with as many as four hundred diggers at a time, and island residents were complaining of vandalism and break ins. By 1989, Savary Island shellfish stocks had collapsed, prompting a closure on the island, which in turn concentrated greater numbers of clam diggers on other beaches that were previously known only to local harvesters. On some of these smaller beaches, the "fishing season" ended up being confined to sixteen days a year.

In 1993, DFO and the provincial fisheries ministry teamed up and approached groups with interests in the clam fishery with a view to developing solutions to the problem. Some of the options were conventional: the introduction of limited-entry licences, such as those that govern most fisheries on the West Coast, and the fashionable remedy of ITQs, individually transferable quotas. Other options were more innovative and community-based, combining elements of traditional fisheries management regimes with "modern" characteristics.

The Sliammon, Klahoose and Sechelt people proposed a system that would limit access to the fishery primarily to local, native harvesters. Local non-native harvesters proposed a system of access rights that was strikingly similar but would guarantee access by non-native local residents with some roots in the industry. By the 1994-95 season, the area's native and non-native harvesters had compromised with a system that ensured equal rights of access to local native and non-native harvesters, under a management board that allowed equal representation by local native, non-native and DFO officials. DFO would retain the right to overrule decisions of the Area C clam management board if conservation was an issue. The management board would participate in the gathering of data and the establishment of overall sustainable harvest levels. A simple formula based on the number of clams an "average" clam digger would harvest in a day would contribute to stock-abundance assessments. Licences would be held by individuals, but they could not be transferred out of the area. Enforcement costs would be covered partly by the harvesters themselves through a 5 per cent levy from the sale of clams.

To its credit, DFO approved the regime. The classic problems that beset the clam fishery—mobility, lack of accountability, overcapacity and overfishing—were solved.

In another area of the province, the Shuswap Nation Fisheries Commission is working jointly with ranchers, loggers and others in Shuswap traditional territory in the development of watershed management planning. By the early 1990s, the Shuswaps were rebuilding some of their ancient fish weirs in the small tributaries of the Thompson River watershed, partly to harvest some salmon and partly to gather run-timing data, conduct salmon counts and assemble stock composition information. Part Coyote, part Mechanism. And at Sooke, on Vancouver Island's southwest coast, the provincial government and DFO were helping the Sooke tribe establish an evolved form of the ancient, live-capture reefnets the people had used there until the canneries came and chased them off their fishing grounds. Instead of cedar-twine reefnets, the Sooke people were building saltwater tidal traps, using log pilings and seine net, inspired partly by Newfoundland's inshore cod-trap fishery. Part Raven, part Come by Chance.

Even so late in the day, the long-term outlook for Canada's West Coast fisheries is not at all doom and gloom. Substantial progress in several areas has already been made, even in the case of the coastal pulp mills that had so badly damaged the coast's delicate marine environment. Years of Greenpeace protests and campaigns by the United Fishermen and Allied Workers Union began to pay off: in the summer of 1995, DFO lifted public health advisories against eating Dungeness and rock crab hepatopancreas from fishing grounds throughout the coast, and 40 per cent of the shellfish closures imposed as a result of dioxin and furan contamination during the 1980s, were lifted. Pollution-control restrictions had produced improvements in dioxin and furan levels well ahead of anybody's expectations. Dioxins in pulp mill emissions had declined by 93 per cent between 1989 and 1993; furans in pulp mill effluent had dropped by 99 per cent over the same period. Progressive new forest legislation was being implemented by the B.C. government, ensuring effective forest cover adjacent to creeks, streams and rivers, and funds raised from timber tenures were being ploughed back into forest renewal initiatives, holding out hope for restored fish habitat.

By 1995, there were even rumours going around about the great kelp forests returning to parts of the coast. The federal government had agreed in the late 1960s to try to restore some of the West Coast's sea otter populations. In the years between 1969 and 1972, a small colony of eighty-nine sea otters, from a stable colony in Alaska, were transplanted to the Bunsby Islands, off Vancouver Island's northwest coast, and there were other minor efforts to bring otters back to their former haunts elsewhere in the vicinity. By the early 1990s, the B.C. coast was home to at least eight hundred sea otters.

It may well be later than we think, but it's not too late.

⚓

CHAPTER VII

Hundreds of Little Jonahs

There is simply nothing in Creation that does not matter.
Our tradition instructs us that this is so, and it is proved to be so,
every day, by our experience. We cannot be improved—in fact we
cannot help but be damaged—by useless or greedy or merely
ignorant destruction of anything.

—Wendell Berry, "The Obligation of Care"

WHEN I WAS A BOY, BYRNE CREEK AROSE IN LITTLE RIVULETS FROM A wooded ravine, and that ravine is still there in 1996, spared the fate that has befallen so many of the Lower Mainland's urban streams. Byrne Creek was badly weakened, but the water course survived pavement, condominiums and industrial landfill partly by luck, partly through Burnaby municipal council's forward-thinking policy of protecting ravines, and partly through the generosity of the local service clubs who saw to it that Ron McLean Park, between Rumble and Marine Drive, would be set aside for coming generations of kids.

Byrne Creek's headwaters now lie mainly underneath Burnaby's supermarkets, suburban streets and apartment buildings. Powerhouse Creek flows underneath Kingsway, just west of its intersection at Edmonds, and joins Byrne Creek somewhere in the storm drains underneath the Edmonds Skytrain station. On these streets, Byrne Creek's cascade can be heard only during the autumn rains, by standing quietly on street corners and listening above the drain grates. Underground, Byrne Creek flows southwest and emerges alive in its ancient canyon, within the remnant forest of Ron McLean Park. Continuing downhill, the creek picks up some more water just below

Marine Drive, where John Matthews Creek trickles down its own ravine between Royal Oak and Gilley avenues. In the industrial area between Marine Drive and Marine Way, Byrne Creek takes in more water, this time from Froggers' Creek, which flows down Burnaby's south slope through another ravine, this one cutting through the suburbs between Royal Oak and Nelson, and then passes between the Chinese Evangelical Church and the Iglesia Ni Cristo onto the flats.

On the flats of the Big Bend, it is different again. The boulders give way to streambed rocks just below the ravines, but once on the flats, the rocks give way to gravel, and Byrne Creek takes in some more water from Gray Creek, which trickles out of the ravine between Nelson and Sussex avenues. From that point the gravel has given way to sand, and the creek flows slowly southwest, through blackberry and sedge and cattails, towards the dyke. In the 1800s, Byrne Creek used to flow east towards the Fraser, through John Woolard's farm, and back then it was called Woolard's Brook. In 1893, Peter Byrne dug a new, straight channel to allow the creek to run due south to the river, and the Gilley Brothers' logging company cut the timber on the south slope, hauling the logs by oxen to Byrne Creek, to float the timber south to the river and then up the Fraser to the New Westminster sawmills. But by the 1990s, Byrne Creek flows southwest, and it empties through flood-control gates into the North Arm of the Fraser River, where it disperses within the waters of thousands more streams from throughout the Fraser basin and continues its life as it began, a small, barely noticeable part of something much bigger. The muddy, roiling North Arm curls gracefully around the Big Bend, taking in Glen-Lyon Creek, Kaymar Creek, Boundary Creek and what little is left of all the other creeks along Vancouver's south slope. The last is Musqueam Creek, and then there is the Strait of Georgia, where the great river itself becomes a small, barely noticeable part of something much bigger.

But salmon notice these things, and sea-run cutthroat trout notice these things, and whatever mystery accounts for it, the olfactory glands, or the molecular memory, or some other intelligence, the fish come up out of the sea after it, and whatever it is that they notice in all of this, they follow it up from the sea into the rivers. However dispersed within the Fraser River one of these tiny creeks might be, they find it, and they trace the underwater course of it, and if they swim too far up the big river they will turn around and find it again, and they will jump right out of the river if that's what they have to do to get to it. They will keep on going, on and on until they are above the sedge and the cattails, and they will head for the gravel, up near the boulders where they began, and they will fight to their death to get there, and they will fight against the very mountains themselves to return to the place they were born. This is the miracle of the salmon we all marvel about. More of a miracle is that they come home to us at all after everything we have done to them. But they do. That's the important part.

They come to Railroad Creek, where a remnant population of coho and chum that somehow survived the Canadian National Railway's excesses continued to return from the sea to spawn, down through the years. Railroad Creek empties into the Fraser River at Dewdney, just east of Mission. At about the time of Hell's Gate, railway crews built up earthworks all along the marshy lowlands behind the river's north bank, not unlike a long, low dam, below the base of the mountains. Crews laid the tracks on top of it all, to keep the line above flood water, putting in a bridge here and there to let the creeks through, punching through a culvert where they had to, but otherwise ignoring the hydrology of the place. Over the years, between the tracks and the mountains, loggers cut down the trees and farmers cleared the land. Countless little creeks were lost, and they were the types of creek chum and coho particularly like—the kind that arise from cold, upwelling springwater at the base of hills and mountains, the kind that flow quickly across gravel-bottom streambeds and empty into a big river nearby, nice and close to the sea. Down through the years, there wasn't much left of the little stream that came to be called Railroad Creek. But there were some chum and coho that spawned there still.

On a winter day in the late 1970s, a spawning pair of coho ventured a few metres past the gravel where they had emerged as eggs. They swam through a culvert that had been constructed through the embankment underneath the railroad tracks. They travelled a short distance up the ditch and found cold, bright water trickling in from a farmer's cow field. Battered and weary, the two coho jumped up and out of the creek, in the direction of the falling water, and landed in the soggy field. Together, they followed the flow of the ankle-deep water to the place where it was welling up through the grass. There, they began to dig with their tails. They unearthed the spawning gravel of their ancestors, at the old source of the creek, and there they spawned and died.

Their offspring returned in great numbers. They returned to Railroad Creek and kept going. They swam through the culvert, up the ditch and into the field, through the grass to the place they were born and began to dig some more. They were followed by chum salmon, the offspring of fish born in Railroad Creek alongside the pioneering cohos' parents.

And so it went. On a cold January afternoon in 1988, biologists Matt Foy and Dave Marshall stood at the edge of a textbook-perfect spawning area, a gravel-covered expanse filled with bubbling, clear, bright springwater, in the middle of a farmer's field. Above the spawning beds, cedar posts hung in mid-air from lines of barbed wire. What had once been a fence wasn't really a fence any more, because the ground had been excavated out from beneath it, bulldozed downstream by the shredded and mangled tails of scores of chum and coho salmon. "I know," Marshall said. "It's hard to believe." A lone male chum salmon, bruised and ragged, swam in the shallows among carcasses of fish that had fought to regain the spawning grounds their ancestors had lost so many decades before. One of the dead fish hung, like vengeance itself, from a strand of barbed wire. "They punched through a cowfield, through that little ditch," Foy said. "The fish themselves actually constructed their own spawning grounds, and every year the run is expanding."

Marshall and Foy are among dozens of fisheries biologists and technicians, many of whom work for the Salmonid Enhancement Program, and many of them senior SEP managers as well, who had little faith left, by the 1980s, in elaborate construction plans for cement hatchery complexes. Watch the fish, these biologists proposed instead. See what they want. Help them get it.

The salmon come to Worth Creek, not far from Railroad Creek, another place where it was obvious to Foy and Marshall what the fish wanted. There were spawners in Worth Creek. SEP technicians bulldozed a cut from the creek to its old source of springwater and emptied truckloads of gravel into the cut. It worked. In 1984, salmon were spawning in it. SEP crews left two dump-truck loads of gravel at the head of the cut and returned in 1987 to make the spawning beds bigger, but the coho and the chum had beaten them to it. The gravel pile was gone. The fish had dug away at the base of it and spread the gravel out neatly, beneath the water, moving it out and downstream. They dug their nests and spawned in it.

A short drive upriver at the mouth of Maria Slough, the long, hot summers and the logged-off mountains cause the slough to get silty and dry up in spots. When the rains come and autumn turns to winter, the coho and chum come back to Hicks Creek, which empties into the slough. The fish clear out the muck every year, opening up the beds again. "It looks like some outfit's gravel operation," Foy said. "Every winter, the coho move in there and dig it away."

Sometimes, it's as though SEP is getting help from the fish, instead of the other way around.

After building a series of costly artificial spawning channels above the Chehalis River, SEP engineers encountered an unforeseen problem. The channels were silting up badly every year, and the whole thing had started to look like a bad idea. Lots of money needed to be spent every year on bulldozers and backhoes to clear deep drifts of sand off the gravel beds. But as the years passed, it became obvious much of the machine engineering was unnecessary. SEP managers cleared out the upper ends of the channel, letting the sand settle downstream. The fish cleared out the rest. "When it settles out, you get it in piles three to four feet deep," said biologist Bruce Shepherd. "The fish move it out themselves. They just dig it out and move it downstream."

In the furthest reaches of the Fraser basin, in the Stuart River system, there are sections of the Tachie River and the Middle River that don't look like any salmon could ever spawn there. As Shepherd described it, "It's muck as deep as you can poke a stick," and the rivers are slow-moving, meandering through flatlands. But over the years, chinook and sockeye salmon have constructed their own spawning platforms that run from one bank to the other, at right angles to the current, with gravel forming the upstream side and mud and sand forming the downstream side, raising the water velocity in the bargain.

So while we have been taught to marvel at the wonder of salmon the great navigator, salmon are also engineers, and they are colonizers, and it took them less than 10,000 years to colonize or recolonize just about every river, stream, creek and ditch between the coast and the Rocky Mountains, much of which remained covered under sheets of ice miles thick, for several centuries after the close of the last great ice age.

It is true that coho were extirpated from Byrne Creek when I was a boy, and there was no evidence of any sea-run cutthroat trout in Byrne Creek by the time I was in my teens. But in the autumn of 1995, I was walking along a trail in Byrne Creek's ravine, just above Marine Drive.

It was Sunday morning and the sunlight fell through the cedar and the fir trees, and there was movement in the creek, in a pool below a boulder. I had a fleeting glimpse of something. It was too hard to see. There were shadows, and whatever it was, it was just a tiny thing, in a quiet reach of the creek, in a tangle of roots, below some overhanging ferns.

"There. Look at that, there. Do you see it?"

Ken Glover, who is seventy-four, and played in these woods and along this creek when he was a boy, was pointing at the water below the boulder. He stepped carefully in his gumboots along the bank, and knelt down. Tony Pletcher, sixty-three, who graduated from my old high school a quarter-century before I did, was upstream a few paces, in gumboots, peering into the shallow water just above the boulder. Bert Richardson, sixty-seven, who has been tracking the fate of Byrne Creek for the past three decades, was wading just downstream. He stopped, and stood still.

Everything was quiet, just the faint hum of traffic coming through the trees. And there it was. A dark flash, the colour of gun metal, not much bigger than a rifle bullet. A tiny fish darted in a quick triangle pattern, then vanished. Then another, and another.

Cutthroat. Probably yearlings.

Glover, Pletcher and Richardson never gave up on Byrne Creek. Over the years, their friends at the Vancouver Angling and Game Association tried to be understanding about their stubbornness, but few were convinced that Byrne Creek was anything more than a lost cause. The meetings of the association's Byrne Creek subcommittee were never exactly standing-room-only affairs.

Over the years, Glover, Pletcher and Richardson have worked with Salmonid Enhancement Program staff, lobbied municipal council, begged and borrowed from the Department of Fisheries and Oceans and the B.C. Environment Ministry and organized teams of schoolchildren from Suncrest, Nelson, Clinton and Stride Avenue schools into bucket brigades every year, releasing coho fry trucked from Kanaka Creek a few miles up the Fraser to the parking lot at Ron McLean Park, where the kids carry the baby fish down into the ravine to watch them swim away.

The coho transplanting began in 1979. There have been returns of jacks, precocious young males who couldn't spawn with a female if they wanted to. And there has been the occasional sighting of adult fish, and it might be that a male and female have paired off, but it's hard to say. Whenever a heavy rain follows a long dry spell, the accumulated toxins in the storm drains produce a lethal "first flush." It is the same all over the Lower Mainland, and what it produces is dead fish. The day we walked Byrne Creek, Chris Savage happened by, walking his dog. Two weeks before, he had seen a dead cutthroat fingerling, followed the foamy water upstream to a cul-de-sac, followed the water halfway down the block. It was just a man innocently washing his car in his driveway, killing fish.

So there are setbacks, but there are victories. Mundy's Towing helped the Byrne Creek volunteers pull tons of truck parts, whole cars, a deep freeze, several refrigerators, several couches, beds and mattresses and dozens of shopping carts out of the creek. Tighter provincial laws and tougher Fisheries Act enforcement have restrained some of the excesses of urban development in the watershed and around the flats. But there remains the question of whether a pair of coho returning to Byrne Creek would survive long enough to spawn, with the water quality as unstable as it is. And there is also the question of whether the baby coho would survive, since insect life has not returned to the creek all that much. But that is life, and you don't give up, Pletcher said. You identify the problems. You work hard to fix the problems.

There are untried remedies, and there are old, tried and true remedies, and there are remedies 'round the world. Maybe there is some scientist somewhere, Pletcher said, working on a methodology that would assist in the restoration of mayflies and crane flies and whatnot—sort of an insect enhancement program. There are experiments that have proven that natural vegetation can be used as a filtration system for toxins, and a simple system of catchment basins might be enough to eliminate the first flush problem. A remedy as low-tech as bags of charcoal dumped in the creek at certain strategic points might serve as a filtering system to maintain water quality, sort of an interim measure until non-toxic detergents are the rule rather than the exception.

"Gumboots," Glover said, walking the trail through the ravine. "More gumboots, fewer computers and word processors, and more gumboots."

"Fish weirs," said Pletcher, himself a biologist. "The Indians should go back to them. Everybody should try using them."

It is these little things that make a difference, in the long run. It is individuals, and small groups of people, working on local initiatives, that end up making a difference. It is not just the fishing industry that has caused the salmon to decline so precipitously. David Salmond Mitchell, the fisheries officer who argued against the dismantling of aboriginal fishing technologies at the turn of the century, understood this only too well. It wasn't just the big things, like Hell's Gate, Mitchell said. It was the little things. It was the hunters who made it their practice to shoot so many eagles and ospreys out of the skies above Shuswap Lake that the salmon-eating coarse fish they preyed upon, like freshwater ling, ran rampant throughout the lakes of the central interior. In the winters that followed the Hell's Gate disaster, Mitchell ventured out into the blizzards, out onto the frozen surface of Shuswap Lake, cut holes through the ice and speared as many lingcod as he could, day in and day out, week in and week out. He enlisted armies of schoolchildren to help him, hauling up the lingcod, cutting open their stomachs and freeing salmon fry, "hundreds of little Jonahs," and Mitchell's contributions to the rebuilding of the great Adams River sockeye runs are incalculable. Maybe he didn't make all that much difference. Or maybe he made all the difference in the world.

Back on Byrne Creek, Ken Glover was taking a rest for a moment in the cool of the ravine. "These things take a long time," Glover said.

The bottom of the ravine was dark, choked with underbrush and alder among the trunks of towering cedar and fir and cottonwood. Byrne Creek riffled quietly at Ken's feet. Every now and then, a dark flash, a tiny fish, darted out and back into the shadows. This is the spot that Glover and the others have been gently herding schoolkids to, year after year. Just upstream from here, the previous winter, Glover had trapped three six-inch, pan-sized cutthroat, just to see if there were any there, and let them go. "When I was a kid, we used to herd the fish back and forth in the creek," Glover said. "What did we know?"

Across the creek, there was a small opening in the side of the sandstone ravine wall, shrouded by broad-leafed maple. I looked closer, and it was as if there was a stone wall back in there, half-covered in the undergrowth. I crossed the creek on the boulders and the stones. It was an old outfall pipe, set in a cement bunker with iron grates covering its dark entrance. Some kid had spray-painted "Hell Hole" above the mouth of the tunnel. It was pitch black inside, as though it went down into the middle of the Earth. I knew this ravine once like it was my own backyard. I didn't remember this.

I crossed back over the creek on the stones, watching the water for movement, and every now and then I caught a little fish out of the corner of my eye—or my mind's eye, anyway.

"It used to come from Dominion Glass," Glover said. "That's what really killed off the creek back then, I think."

And it all started to make more sense. When I looked across the creek, I could see that below the cement wall with Hell Hole gaping out of it there was no underbrush, just exposed stones from the torrents of the years. The drain tunnel led backwards, underground, to the place where the mammoth Dominion Glass factory used to be. I worked there when I was twenty, as far down on the union seniority list as it got. I worked deep within the dark tombs of the place, on the night shift, in winter, shovelling melted, molten glass and huge lumps of silica and caustic fixatives and colouring additives that dropped in great red-hot lumps from the furnaces and the machinery upstairs.

There were no face masks or oxygen tanks or health and safety inspections involved. There were no elaborate settling ponds in the tombs. And it would rain, and eventually all that stuff would sizzle and melt and muddy the pipes, and it would all end up pouring out of Hell Hole. And then the factory shut down, the dump was closed, the condominiums came and the office towers went up.

So it is not good enough to say it was the government, the greedy fishermen, the seiners and the gillnetters and the trollers and the trawlers and the Indians and the poachers, or the seals, or the multinational fishing companies, or the forest companies or the greenhouse effect. I killed Byrne Creek. Who didn't?

It was the jam factory at McPherson and Beresford where I used to beg pop bottles from the women on their lunch break, in the summer, when John Matthews Creek ran red from the berry pulp. It is the man in his driveway on Ewart Avenue washing his car in the fall. It is the antifreeze for his windshield in the winter. It is the bark mulch in the spring around the flowers in your backyard. It is the politician you voted for, the trees you cut down, the grass cuttings you dumped down the hill and the chlorine you put in your pool.

It is as big as Hell's Gate, and it is as small as Hell Hole, and the world gets smaller as you grow older and you learn that things take time, like Ken Glover says. And then you learn that we don't have much time left any more, that we have had all the time in the world.

The world may well be a big place, but it's just as small as Byrne Creek.

Sources

CHAPTER I

Berry, Wendell. "The Obligation of Care." *Sierra,* September/October 1995.

Hillborn, Ray, Ellen Pikitch and Robert Francis. "Current Trends Including Risk and Uncertainty in Stock Assessment and Harvest Decisions." *Canadian Journal of Fisheries and Aquatic Sciences* 50 (1993).

Brown, Lester, et al. *State of the World, 1995: A Worldwatch Institute Report on Progress toward a Sustainable Society.* New York: W. W. Norton and Company, 1995.

Glavin, Terry. *A Ghost in the Water.* Vancouver: Transmontanus/New Star Books, 1994.

Cass, Alan J., Richard J. Beamish and Gordon A. McFarlane. *Lingcod.* Canadian Special Publication of Fisheries and Aquatic Sciences, 109. Ottawa: Department of Fisheries and Oceans, 1990.

Robson, Peter. "The Decline of Abalone: Other Factors." *Westcoast Fisherman,* February 1995.

Hart, John L. *The Pilchard Fishery of British Columbia.* Victoria: Report of the Commissioner of Fisheries, 1933.

Hume, Mark. "Marine Creatures Vanish from Sea and Answers Seem below the Surface." *Vancouver Sun,* March 18, 1995.

Schweigert, J. F. and M. Linekin. *The Georgia and Johnstone Straits Herring Bait Fishery in 1986: Results of a Questionnaire Survey.* Canadian Technical Report, Fisheries and Aquatic Sciences 1721, 1990.

State of the Environment Report for British Columbia, 1993. Victoria: Environment Canada, B.C. Ministry of the Environment, Lands and Parks, 1993.

Forester, Joe and Anne. *B.C. Commercial Fishing History.* Surrey: Hancock House, 1975.

Richards, Laura, Jeff Fargo and John Schnute. "Factors Influencing Bycatch Mortality of Trawl-Caught Pacific Halibut." *North American Journal of Fisheries Management* 15 (1995).

Silver, G. R. and Douglas L. Macleod. *Feasibility of Exploiting Underutilized Species: Biomass Survey.* B.C. Aquaculture Research and Development Council, 1991.

Trawling in British Columbia: Past, Present and Future. Vancouver: Greenpeace Canada, 1995.

Northcote, T. G. and M. D. Burwash. "Fish and Fish Habitats of the Fraser Basin in Dorcey." In *Water in Sustainable Development: Exploring Our Common Future in the Fraser River Basin,* edited by A. H. J. Dorcey and J. R. Griggs. Vancouver: Westwater Research Centre, University of British Columbia, 1991.

Northcote, T. G. and D. Y. Atagi. *Pacific Salmon Abundance Trends in the Fraser River Watershed Compared with Other British Columbia Systems* (in press), 1994.

Northcote, T. G. "The Habitat, Fish and Fisheries of the Fraser River Basin: Are They Sustainable?" In *Proceedings, American Fisheries Society Annual General Meeting.* Halifax, 1994.

Roos, John. *Restoring Fraser River Salmon: A History of the International Pacific Salmon Fisheries Commission, 1937-1985.* Vancouver: Pacific Salmon Commission, 1985.

Ricker, W. E. *Effects of the Fishery and of Obstacles to Migration on the Abundance of Fraser River Sockeye Salmon.* Canadian Technical Report, Fisheries and Aquatic Sciences 1522, 1987.

Riddell, B. E. "Spatial Organization of Pacific Salmon: What to Conserve?" In *Genetic Conservation of Salmonid Fishes*, edited by G. Cloud and G. H. Thorgaard. New York: Plenum Press, 1993.

Beamish, Richard J. and D. R. Boullion. "Pacific Salmon Production Trends in Relation to Climate." *Canadian Journal of Fisheries and Aquatic Sciences* 50 (1993).

Bardach, John. *Sustainable Development of Fisheries.* Honolulu: Environment and Policy Institute, East-West Center, n.d.

Bohn, Glenn. "Hell's Gate Was as Hot as Hades for Sockeye." *Vancouver Sun*, February 21, 1995.

Potential Impacts of Global Warming on Salmon Populations in the Fraser River Watershed. In *Climate Change Digest.* Ottawa: Atmospheric Environment Service, Environment Canada, 1994.

State of the Environment for the Lower Fraser Basin. Ottawa: Environment Canada, 1992.

"So Many Salmon, But So Little." *Scientific American*, May 1995.

CHAPTER II

Norse, Elliott, ed. *Global Marine Biological Diversity: A Strategy for Building Conservation into Decision-Making.*" Washington, D.C.: Island Press, 1993.

Meggs, Geoff. *Salmon: The Decline of the British Columbia Fishery.* Vancouver: Douglas and McIntyre, 1991.

Newell, Dianne, ed. *A Grown Man's Game: The Development of the Pacific Salmon Canning Industry.* Montreal: McGill-Queen's University Press, 1989.

Helgeson, Hans. *Report by Fishery Officer Hans Helgeson, Sessional Paper 22.* 38th annual report of the Department of Marine and Fisheries [for 1905]. Ottawa: 1906.

Copes, Parzival. "An Expanded Salmon Fishery for the Gitksan-Wet'suwet'en in the Upper Skeena Region: Equity Considerations and Management Implications." Unpublished paper, 1988.

Corley-Smith, Peter. *White Bears and Other Curiosities: The First 100 Years of the Royal British Columbia Museum.* Victoria: Ministry of Municipal Affairs, Recreation and Culture, 1989.

McNaughton, Derek. "Fraser Coho in Peril." *Vancouver Sun*, August 2, 1995.

Clark, Colin. "The Economics of Overexploitation: Severe Depletion of Renewable Resources May Result from High Discount Rates Used by Private Exploiters." *Science*, August 1973.

Clark, Colin. "Profit Maximization and the Extinction of Animal Species." *Journal of Political Economy* (1972).

Newell, Dianne. *Tangled Webs of History: Indians and the Law in Canada's Pacific Coast Fisheries.* Toronto: University of Toronto Press, 1993.

Brown, 1995.

Greenpeace, 1995.

Ellis, David. "Analysis of Seine Allocation, Fraser River Sockeye." Unpublished paper, 1993.

Commercial Fishing Licence Report: Seine "A" Licences. Vancouver: Commercial Licence Unit, Department of Fisheries and Oceans, 1993.

ARA Consulting Group. *Seafood Industry Development Planning.* Victoria: Aquaculture and Commercial Fisheries Branch, Fisheries and Food, B.C. Ministry of Agriculture, 1993.

McGoodwin, James R. *Crisis in the World's Fisheries.* Stanford: Stanford University Press, 1990.

Walters, Carl. *Fish on the Line: The Future of Pacific Fisheries.* Vancouver: David Suzuki Foundation, 1995.

Pacific Salmon Commission Run-size Estimation Procedures: An Analysis of the 1994 Shortfall in Escapement of Late-Run Fraser River Sockeye Salmon. Vancouver: Pacific Salmon Commission Technical Report No. 6, May 1995.

Miscellaneous Department of Fisheries and Oceans catch statistics, 1983-91.

CHAPTER III

Mohs, Gordon. *The Upper Sto:lo Indians of British Columbia: An Ethno-Archeological Review.* Prepared for the Alliance of Tribal Councils, 1990.

Bernton, Hal. "Battle for the Deep." *Mother Jones,* August 1994.

Underwood McLellan and Associates, Ltd. *Further production and processing potential for underutilized marine resources of B.C.* Victoria: B.C. Dept. of Economic Development, 1975. Greenpeace, 1995.

Fish Processing Strategic Task Force: A Sectoral Strategy for a Sustainable Fish Processing Industry in British Columbia. Victoria: B.C. Ministry of Agriculture, Fisheries and Food, 1994.

Fisheries Production Statistics of British Columbia, 1991. Victoria: B.C. Ministry of Agriculture, Fisheries and Food, 1993.

Reeves, J. E. and G. S. DiDonato. *Effects of Trawling in Discovery Bay, Washington.* Seattle: Management and Research Division, Washington State Department of Fisheries, 1972.

Alverson, Dayton L., Mark H. Freeberg, Steven A. Murawski and J. G. Pope. *A Global Assessment of Fisheries Bycatch and Discards.* Fisheries Technical Paper 339. New York: United Nations Food and Agriculture Organization, 1993.

Warren, Brad. "The Bycatch Bogey." *Audubon,* May-June, 1994.

Dumping and Discarding in the Groundfish Industry in Atlantic Canada. Fisheries Operations report. Ottawa: Department of Fisheries and Oceans, 1992.

Westrheim, S. J. *The Trawl Fishery in the Strait of Georgia and Vicinity, 1945-1977.* Canadian Manuscript Report of Fisheries and Aquatic Sciences, 1980.

Ketchen, K. S. *An Investigation into the Destruction of Grounds by Otter Trawling Gear.* Progress Report No. 73. Ottawa: Fisheries Research Board of Canada, 1947.

Forrester, C. R. *Life History Information on Some Groundfish Species.* Technical Report No. 105. Ottawa: Fisheries Research Board of Canada, 1969.

Barraclough, W. E. *Decline in the Availability of Brill on the West Coast of Vancouver Island as Associated with a Decline in Recruitment.* Ottawa: Progress Report No. 98. Ottawa: Fisheries Research Board of Canada, 1954.

PSARC Advisory Document 93-3. *Groundfish.* Pacific Stock Assessment Review Committee, 1993.

Richards, Fargo and Schnute, 1995.

Stanley, Red. *A Study of Bottom-trawl Discards in Queen Charlotte Sound and Hecate Strait in 1981 and 1982.* Canadian Technical Report of Fisheries and Aquatic Sciences, No. 1318. Vancouver: Department of Fisheries and Oceans, 1985.

Project Summary—Gear Selectivity. Ottawa: Fisheries Operations, Fishing Industry Services, Department of Fisheries and Oceans, n.d.

Silver, G. R. and Douglas L. Macleod. *Feasibility of Exploiting Underutilized Species—Biomass Survey.* B.C. Aquaculture and Research and Development Council, 1991.

Department of Fisheries and Oceans. Miscellaneous enforcement reports, interdepartmental memoranda and weekly groundfish review reports.

Rice, Jake (head, Marine Fish Division, Science Branch, Department of Fisheries and Oceans). Unclassified memorandum to Frances Dickson, shellfish coordinator, Commercial Fisheries Division, January 8, 1995.

Turris, Bruce R. "Discussion Paper: The Groundfish Trawl Fishery—Where Is It Headed?" Groundfish Management Unit, Department of Fisheries and Oceans, Pacific Region, June 1994. Unpublished.

Beamish, R., F. Bernard, K. Francis, B. Hargreaves, S. McKinnell, L. Margolis and B. Riddell. *A Preliminary Summary of the Impact of the Squid Driftnet Fishery on Salmon, Marine Mammals and Other Marine Animals.* Ottawa: Biological Services Branch, Department of Fisheries and Oceans, 1989.

Bernard, F. R. *Preliminary Report on the Potential Commercial Squid of British Columbia.* Canadian Technical Report of Fisheries and Aquatic Sciences, No. 942. Ottawa: Resource Services Branch, Department of Fisheries and Oceans, 1980.

Department of Fisheries and Oceans. Communique IB-PR-87-03. June 1987.

Burke, William T., Mark Freeburg and Edward L. Miles. "United Nations Resolutions on Driftnet Fishing: An Unsustainable Precedent for High Seas and Coastal Fisheries Management." *Ocean Development and International Law* 25, 1994.

Federal-Provincial Deal to Explore New Fisheries. Joint Canada-British Columbia communique, December 6, 1995.

Trawl Fishers Undertake Experimental Fishing Projects. Westcoast Fisherman, July, 1995.

CHAPTER IV

"Coyote Breaks The Dam." In *Stsepekewlem E—Legends That Teach.* Kamloops: Secwepemc Cultural Education Society, courtesy Shuswap Nation Fisheries Commission, 1994.

Kennedy, Dorothy. *Utilization of Fish by the Chase Indian People of British Columbia.* B.C. Indian Language Project, 1975.

Segal, Howard P. *Technological Utopianism in American Culture.* Chicago: University of Chicago Press, 1985.

Northcote and Burwash, 1991.

Walters, 1995.

Fish Processing Strategic Task Force, 1994.

Glavin, Terry. "Fishplant Jobs Head South—Groundfish Catch On Rise But U.S. Grabs Processing." *Vancouver Sun,* January 3, 1990.

Meffe, Gary K. "An analysis of 'techno-arrogance.'" *Conservation Biology* 6, No.3 (September 1992).

Marsal, Raymond. *Salmon on the Brink—Saving the Wild Chinook in the Georgia Strait.* Vancouver: World Wildlife Fund, 1995.

Pearse, Peter H. *An Assessment of the Salmon Stock Development Program on Canada's Pacific Coast.* Vancouver: Program Review of the Salmonid Enhancement Program, Internal Audit and Evaluation Branch, Department of Fisheries and Oceans, 1994.

Vernon, E. H. *Fraser River Sockeye: The Stocks and Their Enhancement.* A report prepared for the Department of Fisheries and Oceans by Mt. Tolmie Consultants Ltd., 1982.

Pacific Region Salmon Stock Management Plan, Inner South Coast and Fraser River. Vancouver: Department of Fisheries and Oceans, 1988.

Carter, Harry R. and Spencer G. Sealy. *Marbled Murrelet Mortality Due to Gill-net Fishing in Barkley Sound, British Columbia.* Seattle: Proceedings of the Pacific Seabird Group Symposium, 1982.

Graham, C. C. Internal Department of Fisheries and Oceans memorandum to Ed Lochbaum, August 19, 1994.

Weekly Catches by Year of Chinook in Areas 11, 12 and 13. Vancouver: Department of Fisheries and Oceans estimates, August 18, 1994.

Correspondence from Ditidaht First Nation to Fisheries Minister Brian Tobin, June 6, 1995.

Bison, R. G. *The Interception of Steelhead, Chinook and Coho Salmon during Three Commercial Gillnet Openings at Nitinat, 1991.* Report prepared for Ministry of the Environment, Lands and Parks, Fisheries Branch, 1992.

CHAPTER V

Heywood, Robert and Timothy Taylor. *B.C. Commercial Fishing Industry Consultation Document.* Victoria: B.C. Ministry of Agriculture, Fisheries and Food, 1993.

Helin, Calvin. "The Fishing Rights of Canada's First Nations." Unpublished manuscript, 1993.

Harris, Cole. "The Fraser Canyon Encountered." *B.C. Studies* 94 (1992).

Proceedings, Circum-Pacific Pre-History Conference. Seattle: University of Washington, 1989.

Fladmark, Knut. *British Columbia Prehistory.* Archaeological Survey of Canada'. Ottawa: National Museum of Man, 1986.

Mohs, Gordon. *The Upper Sto:lo Indians of British Columbia: An Ethno-Archeological Review.* Prepared for the Alliance of Tribal Councils, February 1990.

Kew, Michael and Julian Griggs. "Native Indians of the Fraser Basin: Towards a Model of Sustainable Resource Use." In *Perspectives on Sustainable Development in Water Management*, edited by A. H. J. Dorcey. Vancouver: Westwater Research Centre, 1991.

Ricker, W. E. *Effects of the Fishery and of Obstacles to Migration on the Abundance of Fraser River Sockeye Salmon (Oncorhynchus nerka).* Canadian Technical Reports 1522, Fisheries and Aquatic Sciences, 1987.

Pearse, Peter H. *Turning the Tide: A New Policy for Canada's Pacific Fisheries.* Vancouver: The Commission on Pacific Fisheries Policy, Final Report, Department of Fisheries and Oceans, 1982.

Hewes, G. W. "Indian Fisheries Productivity in Pre-Contact Times in the Pacific Salmon Area." *Northwest Anthropological Research Notes* 7 (1973).

McMillan, Alan D. *Native Peoples and Cultures of Canada: An Anthropological Overview.* Vancouver: Douglas and McIntyre, 1988.

Kennedy, 1975.

Hayden, Brian, ed. *A Complex Culture of the British Columbia Plateau.* Vancouver: University of British Columbia Press, 1992.

Tyhurst, Robert. "The Chilcotin: An Ethnographic History." Thesis, University of British Columbia, 1984.

Tobey, Margaret. *The Carrier. Smithsonian Handbook of North American Indians,* Volume Six (Sub-Arctic). Washington, D.C.: Smithsonian Institution, 1981.

Lamb, W. Kaye. *The Letters and Journals of Simon Fraser, 1806-1808.* Toronto: Macmillan, 1960.

Teit, James. *The Thompson Indians of British Columbia. Memoirs of the American Museum of Natural History, Volume 1, Part IV,* The Jesup North Pacific Expedition, edited by Franz Boas, New York, 1900.

Duff, Wilson. *The Upper Sto:lo Indians of the Fraser River, British Columbia.* Anthropology in British Columbia, Memoir Number 1. Victoria: Royal British Columbia Museum, 1952.

Roos, 1985.

Ware, Reuben M. *Five Issues, Five Battlegrounds—An Introduction to the History of Indian Fishing in British Columbia, 1850-1930.* Coqualeetza Education Training Centre, 1983.

Mitchell, Sam. *Lillooet Stories. Sound Heritage* 6, No. 1, edited by Dorothy Kennedy and Randy Bouchard. Victoria: B.C. Cultural Services Branch, 1977.

Annual Report. Department of Fisheries, Canada Sessional Papers, No. 22, 1914.

B.C. Fisheries Inspector Annual Report for 1887, Canada Sessional Papers No. 16.

Annual Report [for 1893], Department of Fisheries, Canada Sessional Papers, No. 11 (1894).

Mitchell, D. S. *A Story of the Fraser River's Great Sockeye Runs And Their Loss (Being part of a local history written for my neighbours of the Shuswaps).* Unpublished manuscript, 1925.

"Indians Wiping Out Sockeyes." *Province,* November 19, 1904.

Newell, 1993.

Lillooet Tribe communique, June 8, 1988.

Bijsterveld, L. and M. James. *The Indian Food Fishery in the Pacific Region: Salmon Catches, 1951 to 1984.* Canadian Data Report of Fisheries and Aquatic Sciences, No. 627. Vancouver: Department of Fisheries and Oceans Field Services Branch, 1982.

Macdonald, A. L. *The Indian Food Fishery of the Fraser River: 1991 Summary.* Canadian Data Report of Fisheries and Aquatic Sciences, No. 876. Vancouver: Fisheries Branch, Department of Fisheries and Oceans, 1992.

Department of Fisheries and Oceans. Miscellaneious Indian Food Fishery summaries.

Aboriginal Fisheries Strategy Agreement. Miscellaneous allocation tables.

Sparrow v. Her Majesty the Queen, Supreme Court of Canada, 1990.

James, M. D. *Historic and Present Native Participation in Pacific Coast Commercial Fisheries.* Vancouver: Program Planning and Economics Branch, Department of Fisheries and Oceans, 1984.

James, M. D. *Review of Indian Participation in the 1985 Commercial Salmon Fishery.* Vancouver: Program Planning and Economics Branch, Department of Fisheries and Oceans, 1987.

Cullen, Mary K. *The History of Fort Langley: 1827-96.* Exhibit 29, Tab 5, Her Majesty the Queen v. Dorothy Marie Van Der Pete. nd.

Yaremchuk, Gerald (chief of planning and implementation, aboriginal affairs branch, Department of Fisheries and Oceans). Memorandum to Gus Jaltema, treaty negotiations, aboriginal fisheries, Pacific region, Department of Fisheries and Oceans, March 8, 1995.

Ellis, D. W. "Commercial Fishery Catch Distribution Estimates." Unpublished paper, 1993.

Tables, Pacific Salmon Commission Sockeye Review, September 29, 1995.

Fraser, John, et al. *Fraser River Sockeye 1994—Problems and Discrepancies.* Report of the Fraser River Sockeye Public Review Board. Vancouver: Department of Fisheries and Oceans, 1995.

Catches of Sockeye Salmon in the Fraser River, 1994. Report to the Fraser Sockeye Public Review Board, Fraser River Sockeye Salmon Management Review Team (In-River Catch Estimation Working Group). Vancouver: Department of Fisheries and Oceans, 1995.

Pacific Salmon Commission Run-size Estimation Procedures: An Analysis of the 1994 Shortfall in Escapement of Late-Run Fraser River Sockeye Salmon. Technical Report Number 6. Vancouver: Pacific Salmon Commission, 1995.

Pearse, Peter and Peter Larkin. *Managing Salmon in the Fraser.* Vancouver: Department of Fisheries and Oceans, 1992.

Glavin, Terry. *Rebuilding Sto:lo Fisheries Law: Report of the Community Consultation Process.* Chilliwack: Lower Fraser Fishing Authority, 1993.

Department of Fisheries and Oceans, Management Biology Unit. Miscellaneous Sockeye Allocations and Catches for Fraser River First Nations, 1995.

Gordon, Doug. *Making Waves* 4, No. 3 (1993).

Walker, Michael. "Sports Fishing Better, Economically." *Vancouver Sun,* August 24, 1995.

CHAPTER VI

Pinkerton, Evelyn and Martin Weinstein. *Fisheries That Work—Sustainability through Community-Based Management.* Vancouver: David Suzuki Foundation, 1995.

Greenpeace, 1995.

Greer, Alan. *Local Salmon Management: A Proposal for Co-operative, Community-Based Management of Canada's Pacific Salmon Resource.* Discussion Paper. Vancouver: Program Planning and Economics Branch, Department of Fisheries and Oceans, Pacific Region, 1993.

Short, K. "Japanese Fishery Resource Management under the Rights System: A Case Study from Hokkaido." In *A Sea of Small Boats,* edited by J. Cordell. Cambridge: Cultural Survival Press, 1989.

CHAPTER VII

Berry, 1995.

Brown, 1995.

McGoodwin, 1990.

Glavin, Terry. "Fish Grow in a Farmer's Field." *Vancouver Sun,* January 30, 1988.

Glavin, Terry. "Story of Fish-Filled River Was Never Published." *Vancouver Sun,* June 5, 1991.

Index

Aboriginal Fisheries Commission, 76
Aboriginal Fisheries Strategy, 43, 74, 115-23, 130, 137, 145, 155. See also Native fishery
Aboriginal fishing rights. See Native fishery
Adams River salmon run, 121, 126
Adkins, Bruce, 12
Aluminum Company of Canada, 20, 78
Anglo-British Company, 118
Arctic Harvester (trawler), 52, 53
Area C (clam harvest area), 159, 160
Atagi, Dana, 14

Babine Lake, 150
Babine people, 26, 28, 29-30
Bailey, Rick, 148
B.C. Fisheries Commission, 116
B.C. Packers Ltd., 27, 41, 53, 80, 82, 117-18, 120
B.C. Treaty Commission, 123
Beach-seine fishery, 148-49, 151
Birch, Edward, 84, 85
Bison, Rob, 91
Black cod, 55, 57, 60-61
Boetscher-Ignace, Marianne, 104, 105
Boldt decision, 43
Bouchard, Randy, 103, 105
Brodeur, Louis, 28
Burwash, M. D., 14
"Buyback," of fishing vessels, 48, 49
Bycatch. See Catch
Byrne Creek, 1-2, 162, 163, 168-70

Canadian Fishing Company Ltd., 41, 42, 53, 80, 118
Canneries, 14-15, 26-30, 81, 106-8, 110
Carrier people, 99, 103, 104, 124-25
Catch: B.C. annual, 22; incidental, mortality of, 7, 37; incidental, of driftnet fishery, 71-72; incidental, of gillnet fishery, 87; incidental, of trawl fishery, 13-14, 55, 56-57; incidental, of troll fishery, 89-90; individually transferable quotas (ITQs), 140, 146; international annual, 22-23, 36; juvenile and undersized fish, 89-90; pre-contact native, on Fraser River, 102, 103; rise in, after licence reduction, 49; use for non-human consumption, 37
Chehalis River, 166
Chinook salmon: allocation of Alaskan fishers, 5; conservation efforts, 6, 111; decline of stocks, 14, 15, 30-31, 79, 84; as incidental catch, 63, 87, 88, 89, 147; non-survival at sea, 18; as non-target species of commercial fishery, 33-34; as preferred native fish, 100, 102, 110-11; and sports fishery, 5, 111, 147
Chisholm, Brian, 98
Chum salmon: decline of stocks, 14, 79; as

incidental catch, 90-91, 147; Nitinat fishery, 90-91; preferred by natives for smoking, 102; processing, and price variations, 143; in Railroad Creek, 164-65; value, and management costs, 145
Clark, Colin, 34
Cod, Atlantic, decline of stocks, 3, 9
Cod, Pacific, 53
Coho salmon: in Byrne Creek, 2; conservation efforts, 150; and dams, effect on, 11; decline of stocks, 6-7, 14, 15, 30-31, 84; as incidental catch, 7, 87, 88-89, 91; non-survival at sea, 18; as non-target species of commercial fishery, 33-34; in Railroad Creek, 164-65; as traditional native catch, 100
Columbia River, 10, 18
Commercial fishery: and Aboriginal Fisheries Strategy, 145; B.C. annual catch, 22; catch decline, economic effects of, 125-26; compared with community-based fisheries, 37-39; concentrated ownership of, 41-42; cooperation with native bands, 90-91; cooperation with Skeena River interest groups, 151; depletion of resources by, 39; disregard for diversity of salmon stocks, 33-34; investment in salmon farms, 39; mobility of, 46-47, 50; native participation in, 118; non-selective fishing techniques, 30-31; objection to native fish sales, 117; offshore, 38; opposition to Aboriginal Fisheries Strategy, 119-21; overcapacity of, 31-32, 36, 39-40, 47-50, 138-39; preference for sockeye, 7, 110, 112; privatization, arguments against, 129, 138-39; protests against, by fishers, 127-31; subsidization of, 36; and sustainability, 33, 46, 81. See also Fishery; Gillnet fishery; Native fishery; Trawl fishery; Troll fishery; names of fish species
Community-based fisheries, 128-61; Alaska, 151-53; benefits of, 141; and Community Development Quotas (CDQs), 146; compared with commercial fishery, 37-39; inshore, 38; Japan, 153-55; prevalence of, 37; rejected as option by Fish Processing Strategic Task Force Report, 82-83; replacement of, by commercial fishery, 36; Skeena River area, 149-51; and sustainability, 31-32, 38; traditional native fishery, 101-2
Community Development Quotas (CDQs), 146
Cooperatives, 117-18, 153-55, 157-58
Copes, Parzival, 29-30, 31
Coquitlam people, 100, 117
Crab, 55, 57, 61, 161
Crosbie, John, 45, 115, 116, 120